D0935646

Mechanical Engineering Series

Frederick F. Ling
Series Editor

Springer
New York
Berlin
Heidelberg
Hong Kong
London
Milan
Paris
Tokyo

Mechanical Engineering Series

J. Angeles, **Fundamentals of Robotic Mechanical Systems: Theory, Methods, and Algorithms, 2nd ed.**

P. Basu, C. Kefa, and L. Jestin, **Boilers and Burners: Design and Theory**

J.M. Berthelot, **Composite Materials: Mechanical Behavior and Structural Analysis**

I.J. Busch-Vishniac, **Electromechanical Sensors and Actuators**

J. Chakrabarty, **Applied Plasticity**

G. Chryssolouris, **Laser Machining: Theory and Practice**

V.N. Constantinescu, **Laminar Viscous Flow**

G.A. Costello, **Theory of Wire Rope, 2nd ed.**

K. Czolczynski, **Rotordynamics of Gas-Lubricated Journal Bearing Systems**

M.S. Darlow, **Balancing of High-Speed Machinery**

J.F. Doyle, **Nonlinear Analysis of Thin-Walled Structures: Statics, Dynamics, and Stability**

J.F. Doyle, **Wave Propagation in Structures: Spectral Analysis Using Fast Discrete Fourier Transforms, 2nd ed.**

P.A. Engel, **Structural Analysis of Printed Circuit Board Systems**

A.C. Fischer-Cripps, **Introduction to Contact Mechanics**

A.C. Fischer-Cripps, **Nanoindentation, 2nd ed.**

J. García de Jalón and E. Bayo, **Kinematic and Dynamic Simulation of Multibody Systems: The Real-Time Challenge**

W.K. Gawronski, **Advanced Structural Dynamics and Active Control of Structures**

W.K. Gawronski, **Dynamics and Control of Structures: A Modal Approach**

K.C. Gupta, **Mechanics and Control of Robots**

J. Ida and J.P.A. Bastos, **Electromagnetics and Calculations of Fields**

M. Kaviany, **Principles of Convective Heat Transfer, 2nd ed.**

M. Kaviany, **Principles of Heat Transfer in Porous Media, 2nd ed.**

(continued after index)

Anthony C. Fischer-Cripps

Nanoindentation

Second Edition

With 103 Figures

Anthony C. Fischer-Cripps
CSIRO
Bradfield Road, West Lindfield
Lindfield NSW 2087, Australia
Tony.Cripps@csiro.au

Series Editor
Frederick F. Ling
Ernest F. Gloyna Regents Chair
in Engineering
Department of Mechanical Engineering
The University of Texas at Austin
Austin, TX 78712-1063, USA
and
William Howard Hart Professor Emeritus
Department of Mechanical Engineering,
Aeronautical Engineering and Mechanics
Rensselaer Polytechnic Institute
Troy, NY 12180-3590, USA

ISBN 0-387-22045-3 Printed on acid-free paper.

Printed in the United States of America. (SB/BP)

9 8 7 6 5 4 3 2 1 SPIN 10963654

Springer-Verlag is a part of *Springer Science+Business Media*

springeronline.com

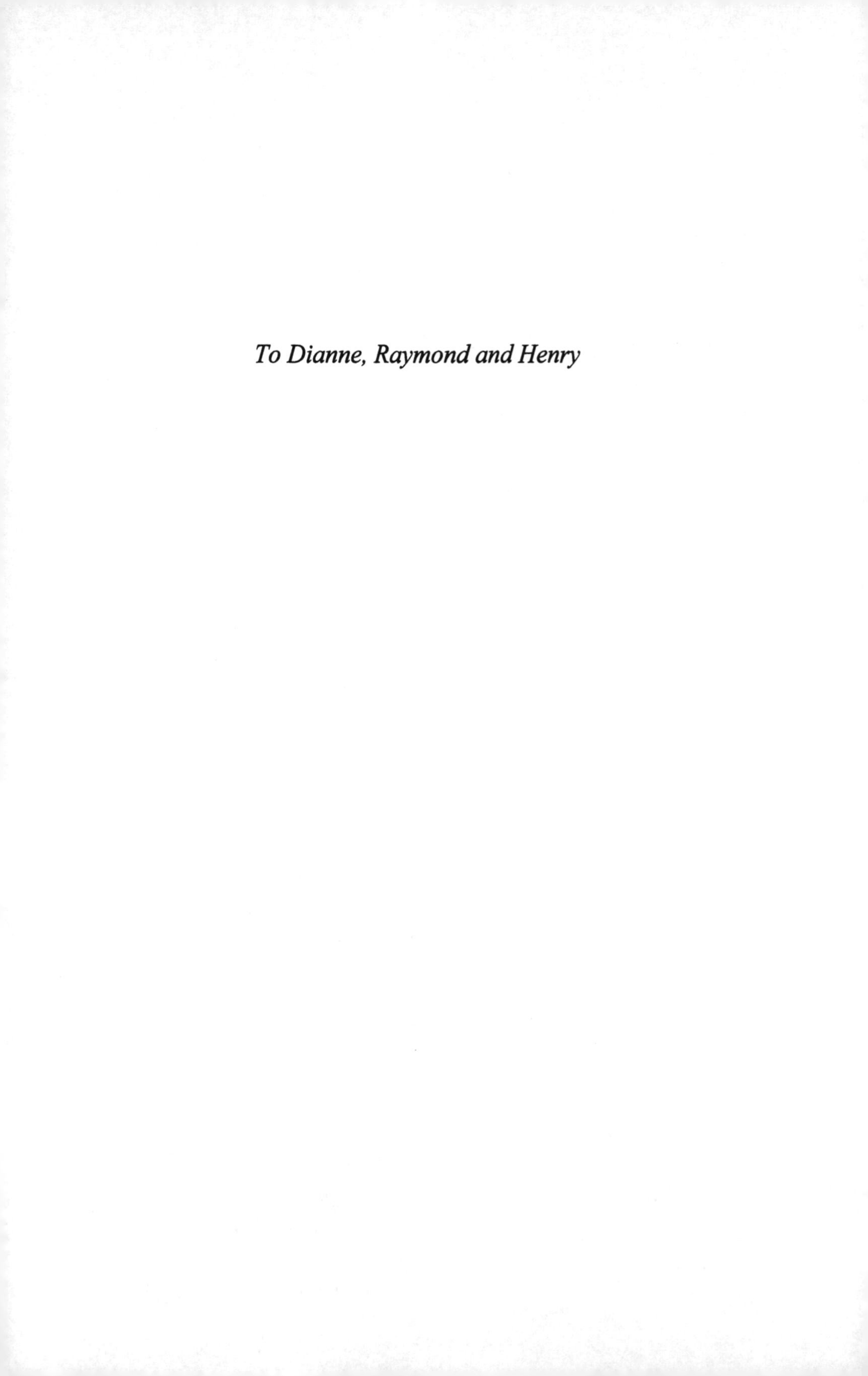

To Dianne, Raymond and Henry

Mechanical Engineering Series

Frederick F. Ling
Series Editor

The Mechanical Engineering Series features graduate texts and research monographs to address the need for information in contemporary mechanical engineering, including areas of concentration of applied mechanics, biomechanics, computational mechanics, dynamical systems and control, energetics, mechanics of materials, processing, production systems, thermal science, and tribology.

Series Preface

Mechanical engineering, an engineering discipline forged and shaped by the needs of the industrial revolution, is once again asked to do its substantial share in the call for industrial renewal. The general call is urgent as we face profound issues of productivity and competitiveness that require engineering solutions. The Mechanical Engineering Series features graduate texts and research monographs intended to address the need for information in contemporary areas of mechanical engineering.

The series is conceived as a comprehensive one that covers a broad range of concentrations important to mechanical engineering graduate education and research. We are fortunate to have a distinguished roster of consulting editors on the advisory board, each an expert in one of the areas of concentration. The names of the consulting editors are listed on the facing page of this volume. The areas of concentration are applied mechanics, biomechanics, computational mechanics, dynamic systems and control, energetics, mechanics of materials, processing, production systems, thermal science, and tribology.

New York, New York — Frederick F. Ling

Preface

There has been considerable interest in the last two decades in the mechanical characterisation of thin film systems and small volumes of material using depth-sensing indentation tests utilising either spherical or pyramidal indenters. Usually, the principal goal of such testing is to obtain values for elastic modulus and hardness of the specimen material from experimental readings of indenter load and depth of penetration. The forces involved are usually in the millinewton range and are measured with a resolution of a few nanonewtons. The depths of penetration are in the order of nanometres, hence the term "nanoindentation."

This second edition of Nanoindentation corrects many errors in the first edition while at the same time, includes an account of the most recent research along with a large amount of helpful information about nanoindentation instruments. The book is intended for those who are entering the field for the first time and to act as a reference for those already conversant with the technique.

In preparing this book, I was encouraged and assisted by many friends and colleagues. Particular thanks to Trevor Bell, Andy Bushby, Avi Bendavid, Alec Bendeli, Robert Bolster, Yang-Tse Cheng, Christophe Comte, Peter Cusack, John Field, Asa Jamting, Nigel Jennett, Brian Lawn, Boon Lim, Alfonso Ngan, Darien Northcote, Nicholas Randall, Paul Rusconi, Doug Smith, Jim Smith, Eric Thwaite, Yvonne Wilson, Oden Warren, and Thomas Wyrobek for their advice and assistance. I thank the following companies for their important contributions: CSM Instruments, Hysitron Inc., and Micro Materials Ltd. I gratefully acknowledge the support of the CSIRO Division of Telecommunications and Industrial Physics and, in particular, Ken Hews-Taylor who supported the UMIS instrument for many years in his management portfolio, the staff of the library, and the Chief of the Division for his permission to use the many figures that appear in this book. I also thank the many authors and colleagues who publish in this field from whose work I have drawn and without which this book would not be possible. Finally, I thank the editorial and production team at Springer-Verlag New York, Inc., for their very professional and helpful approach to the whole publication process.

Lindfield, Australia

Anthony C. Fischer-Cripps

Contents

List of Symbols

α	cone semi-angle, geometry correction factor for Knoop indenter analysis, surface roughness parameter, thin film hardness parameter, buckling parameter
β	indenter cone inclination angle, indenter geometry shape factor
ϕ	phase angle between force and depth in oscillatory indentation tests
δ	distance of mutual approach between indenter and specimen
ε	strain
γ	half of the total energy required to separate two surfaces
Γ	gamma function
η	coefficient of viscosity
θ	angle
ρ	number density of molecules
σ	normal stress
σ_I	indentation stress
σ_r	residual stress
σ_s	maximum asperity height
σ_z	normal pressure underneath the indenter
τ	shear stress
μ	coefficient of friction
ν	Poisson's ratio
ω	frequency
a	radius of circle of contact, constant for linear fit
A	contact area, constant for P vs h relationship
a_c	radius of circle of contact at transition from elastic to plastic deformation with spherical indenter
A_f	portion of contact area carried by film
A_i	area of contact that would be obtained for an ideal indenter at a particular penetration depth
a_o	contact radius obtained for smooth surfaces
A_p	projected contact area
A_s	portion of contact area carried by substrate
b	length of the short diagonal of the residual impression made by a Knoop indenter, Burgers vector, constant for linear fit
C	constraint factor, coefficients for area function expansion
C_0	size of plastic zone

C_f load frame compliance
d length of diagonal of residual impression, diameter of residual impression, length of long side of impression from a Knoop indenter
D diameter of spherical indenter, damping factor
E elastic modulus
E^* combined or reduced elastic modulus
E_{eff} effective modulus of film and substrate combination
E_f film modulus
E_s substrate modulus
F force
F_L lateral force (normal to scratch)
F_N normal force
F_T tangential force (parallel to scratch)
G shear modulus, storage modulus, loss modulus
h indentation depth
H hardness
h^* characteristic length for depth dependence on hardness
h_a depth of circle of contact measured from specimen free surface
h_e elastic depth of penetration for unloading
H_{eff} effective hardness of film–substrate combination
H_f film hardness
h_i initial penetration depth
h_o amplitude of oscillatory depth reading
H_o hardness measured without presence of dislocations
h_p depth of circle of contact measured from maximum depth h_t (the plastic depth)
h_r depth of residual impression
h_{rp} plastic depth of penetration for an equivalent punch
h_s penetration depth at unloading force P_s, depth at which spherical indenter tip meets conical support measured from indenter tip
H_s substrate hardness
h_t total indentation depth measured from specimen free surface
I_o weighting function for thin film analysis
K constant for determining initial penetration depth, Boltzmann's constant, bulk modulus, intercept correction factor, coefficient for stress–strain response in uniaxial plastic regime
K_c fracture toughness
K_s stiffness of indenter support springs
L, l length or distance
m mass of oscillating components, power law exponent that describes the form of the loading and unloading curves
n Meyer's index, slope of logarithmic method of determining h_I
P indenter load (force), hydrostatic pressure
P_c critical load at onset of plastic deformation with spherical indenter, pull-off load due to adhesive forces

P_I	indenter load at initial penetration
p_m	mean contact pressure
p_o	amplitude of oscillatory force
P_s	indenter load at partial unload
R	spherical indenter radius
r	radial distance measured from axis of symmetry
R^+	equivalent rigid indenter radius for contact involving a deformable indenter of radius R
R_i	radius of indenter
R_o	initial radius of curvature
R_r	radius of curvature of residual impression
S	contact stiffness dP/dh
S_L	slope of loading curve
S_U	slope of unloading curve
T	temperature, interfacial shear strength
t	time, film thickness
t_c	coating (film) thickness
t_f	film thickness
t_s	substrate thickness
U	energy, work
u_z	displacement
V	volume
w	interaction potential
W	work
x	strain-hardening exponent
Y	yield stress
Y_f	yield stress of film
Y_s	yield stress of substrate
z_o	equilibrium spacing in the Lennard-Jones potential

Introduction

Indentation testing is a simple method that consists essentially of touching the material of interest whose mechanical properties such as elastic modulus and hardness are unknown with another material whose properties are known. The technique has its origins in Mohs' hardness scale of 1822 in which materials that are able to leave a permanent scratch in another were ranked harder material with diamond assigned the maximum value of 10 on the scale. The establishment of the Brinell, Knoop, Vickers, and Rockwell tests all follow from a refinement of the method of indenting one material with another. Nanoindentation is simply an indentation test in which the length scale of the penetration is measured in nanometres (10^{-9} m) rather than microns (10^{-6} m) or millimetres (10^{-3} m), the latter being common in conventional hardness tests. Apart from the displacement scale involved, the distinguishing feature of most nanoindentation testing is the indirect measurement of the contact area — that is, the area of contact between the indenter and the specimen. In conventional indentation tests, the area of contact is calculated from direct measurements of the dimensions of the residual impression left in the specimen surface upon the removal of load. In nanoindentation tests, the size of the residual impression is of the order of microns and too small to be conveniently measured directly. Thus, it is customary to determine the area of contact by measuring the depth of penetration of the indenter into the specimen surface. This, together with the known geometry of the indenter, provides an indirect measurement of contact area at full load. For this reason, nanoindentation testing can be considered a special case of the more general terms: depth-sensing indentation (DSI) or instrumented indentation testing (IIT).

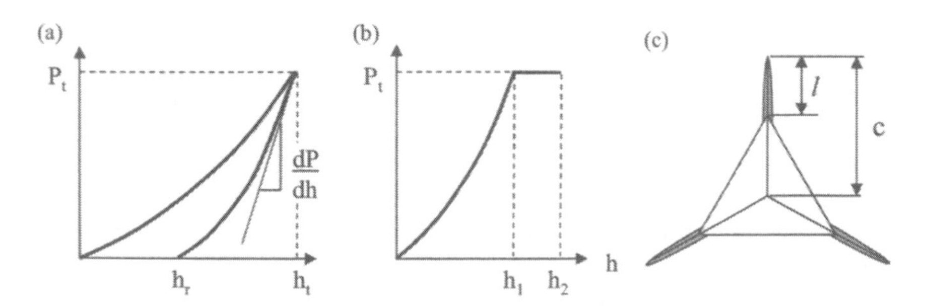

Fig. 1 Load-displacement curves for (a) an elastic plastic solid and (b) a viscoelastic solid for a spherical indenter and (c) cracks emanating from the corners of the residual impression in a brittle material.

It is not only hardness that is of interest to materials scientists. Indentation techniques can also be used to calculate elastic modulus, strain-hardening exponent, fracture toughness (for brittle materials), and viscoelastic properties. How can such a wide variety of properties be extracted from such a simple test, which, in many respects, can be considered a "non-destructive" test method? Consider the load-displacement response shown in Fig. 1. This type of data is obtained when an indenter, shaped as a sphere, is placed into contact with the flat surface of the specimen with a steadily increasing load. Both load and depth of penetration are recorded at each load increment (ultimately providing a measure of modulus and hardness as a function of depth beneath the surface). Following the attainment of the maximum load, in the material shown in Fig. 1 (a), the load is steadily removed and the penetration depth recorded. The loading part of the indentation cycle may consist of an initial elastic contact, followed by plastic flow, or yield, within the specimen at higher loads. Upon unloading, if yield has occurred, the load-displacement data follow a different path until at zero applied load, a residual impression is left in the specimen surface. The maximum depth of penetration for a particular load, together with the slope of the unloading curve measured at the tangent to the data point at maximum load, lead to a measure of both hardness and elastic modulus of the specimen material. In some cases, it is possible to measure elastic modulus from not only the unloading portion, but also the loading portion of the curve. For a viscoelastic material, the relationship between load and depth of penetration is not linearly dependent. That is, for a given load, the resulting depth of penetration may depend upon the rate of application of load as well as the magnitude of the load itself. For such materials, the indentation test will be accompanied by "creep," and this manifests itself as a change in depth for a constant applied load as shown in Fig. 1 (b). An analysis of the creep portion of the load-displacement response yields quantitative information about the elastic "solid-like" properties of the specimen, and also the "fluid-like" or "out-of-phase" components of the specimen properties. In brittle materials, cracking of the specimen may occur, especially when using a pyramidal indenter such as the three-sided Berkovich or the four-sided Vickers indenter. As shown in Fig. 1 (c), the length of the crack, which often begins at the corners of the indentation impression, can be used to calculate the fracture toughness of the specimen material.

More advanced methods can be employed to study residual stresses in thin films, the properties of materials at high temperatures, scratch resistance and film adhesion, and, in some cases, van der Waals type surface forces. In this book, all these issues are examined and reported beginning with a description of the method of test and the basis upon which the analysis is founded. Later chapters deal with the various corrections required to account for a number of instrumental and materials related effects that are a source of error in the measurement, theoretical aspects behind the constitutive laws that relate the mechanical properties to the measurement quantities, recent attempts at formulating an international standard for nanoindentation, examples of applications, and a brief description of commercially available instruments.

Chapter 1
Contact Mechanics

1.1 Introduction

There has been considerable recent interest in the mechanical characterisation of thin film systems and small volumes of material using depth-sensing indentation tests with either spherical or pyramidal indenters. Usually, the principal goal of such testing is to extract elastic modulus and hardness of the specimen material from experimental readings of indenter load and depth of penetration. These readings give an indirect measure of the area of contact at full load, from which the mean contact pressure, and thus hardness, may be estimated. The test procedure, for both spheres and pyramidal indenters, usually involves an elastic–plastic loading sequence followed by an unloading. The validity of the results for hardness and modulus depends largely upon the analysis procedure used to process the raw data. Such procedures are concerned not only with the extraction of modulus and hardness, but also with correcting the raw data for various systematic errors that have been identified for this type of testing. The forces involved are usually in the millinewton (10^{-3} N) range and are measured with a resolution of a few nanonewtons (10^{-9} N). The depths of penetration are on the order of microns with a resolution of less than a nanometre (10^{-9} m). In this chapter, the general principles of elastic and elastic–plastic contact and how these relate to indentations at the nanometre scale are considered.

1.2 Elastic Contact

The stresses and deflections arising from the contact between two elastic solids are of particular interest to those undertaking indentation testing. The most well-known scenario is the contact between a rigid sphere and a flat surface as shown in Fig. 1.1.

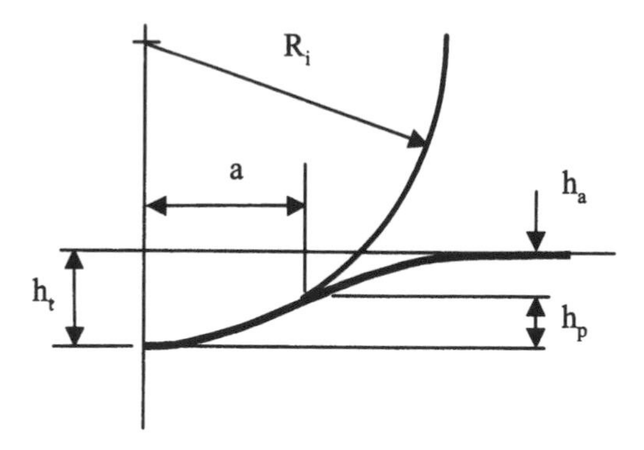

Fig. 1.1 Schematic of contact between a rigid indenter and a flat specimen with modulus E. The radius of the circle of contact is a, and the total depth of penetration is h_t. h_a is the depth of the circle of contact from the specimen free surface, and h_p is the distance from the bottom of the contact to the contact circle.

Hertz[1,2] found that the radius of the circle of contact a is related to the indenter load P, the indenter radius R, and the elastic properties of the contacting materials by:

$$a^3 = \frac{3}{4}\frac{PR}{E^*} \tag{1.2a}$$

The quantity E^* combines the modulus of the indenter and the specimen and is given by[3]:

$$\frac{1}{E^*} = \frac{\left(1-\nu^2\right)}{E} + \frac{\left(1-\nu'^2\right)}{E'} \tag{1.2b}$$

where the primed terms apply to the indenter properties. E^* is often referred to as the "reduced modulus" or "combined modulus" of the system. If both contacting bodies have a curvature, then R in the above equations is their relative radii given by:

$$\frac{1}{R} = \frac{1}{R_1} + \frac{1}{R_2} \tag{1.2c}$$

In Eq. 1.2c the radius of the indenter is set to be positive always, and the radius of the specimen to be positive if its center of curvature is on the opposite side of the lines of contact between the two bodies.

It is important to realize that the deformations at the contact are localized and the Hertz equations are concerned with these and not the bulk deformations and stresses associated with the method of support of the contacting bodies. The deflection h of the original free surface in the vicinity of the indenter is given by:

$$h = \frac{1}{E^*}\frac{3}{2}\frac{P}{4a}\left(2 - \frac{r^2}{a^2}\right) \qquad r \leq a \tag{1.2d}$$

It can be easily shown from Eq. 1.2d that the depth of the circle of contact beneath the specimen free surface is half of the total elastic displacement. That is, the distance from the specimen free surface to the depth of the radius of the circle of contact at full load is $h_a = h_p = h_t/2$:

The distance of mutual approach of distant points in the indenter and specimen is calculated from

$$\delta^3 = \left(\frac{3}{4E^*}\right)^2 \frac{P^2}{R} \tag{1.2e}$$

Substituting Eq. 1.2d into 1.2a, the distance of mutual approach is expressed as:

$$\delta = \frac{a^2}{R} \tag{1.2f}$$

For the case of a non-rigid indenter, if the specimen is assigned a modulus of E^*, then the contact can be viewed as taking place between a rigid indenter of radius R. δ in Eq. 1.2e becomes the total depth of penetration h_t beneath the specimen free surface. Rearranging Eq. 1.2e slightly, we obtain:

$$P = \frac{4}{3}E^* R^{1/2} h_t^{3/2} \tag{1.2g}$$

Although the substitution of E^* for the specimen modulus and the associated assumption of a rigid indenter of radius R might satisfy the contact mechanics of the situation by Eqs. 1.2a to 1.2g, it should be realized that for the case of a non-rigid indenter, the actual deformation experienced by the specimen is that obtained with a contact with a rigid indenter of a larger radius R^+ as shown in Fig. 1.2. This larger radius may be computed using Eq. 1.2a with E' in Eq. 1.2b set as for a rigid indenter. In terms of the radius of the contact circle a, the equivalent rigid indenter radius is given by[4]:

$$R^+ = \frac{4a^3 E}{3\left(1 - \nu^2\right)P} \tag{1.2h}$$

There have been some concerns raised in the literature[5] about the validity of the use of the combined modulus in these equations but these have been shown to be invalid.[4,6] Even if the deformation of the indenter is accounted for, the result, correctly interpreted, is equivalent to a rigid indenter in contact with a compliant specimen.

The mean contact pressure, p_m, is given by the indenter load divided by the contact area and is a useful normalizing parameter, which has the additional virtue of having actual physical significance.

$$p_m = \frac{P}{\pi a^2} \tag{1.2i}$$

Combining Eqs. 1.2a and Eq. 1.2i, we obtain:

$$p_m = \left(\frac{4E^*}{3\pi}\right)\frac{a}{R} \tag{1.2j}$$

The mean contact pressure is often referred to as the "indentation stress" and the quantity a/R as the "indentation strain." This functional relationship between p_m and a/R foreshadows the existence of a stress–strain response similar in nature to that more commonly obtained from conventional uniaxial tension and compression tests. In both cases, a fully elastic condition yields a linear response. However, owing to the localized nature of the stress field, an *indentation* stress–strain relationship yields valuable information about the elastic–plastic properties of the test material that is not generally available from uniaxial tension and compression tests.

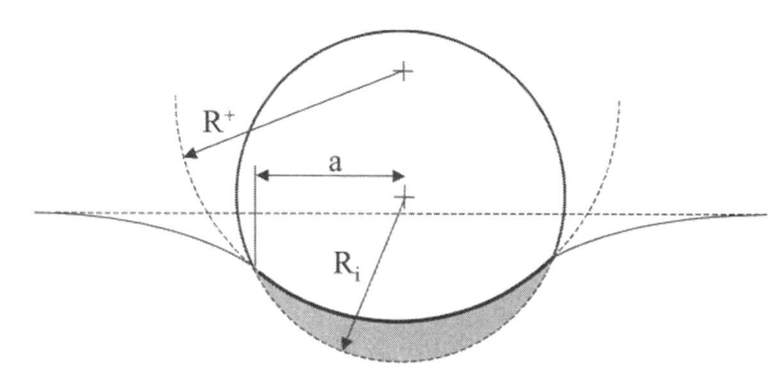

Fig. 1.2 Contact between a non-rigid indenter and the flat surface of a specimen with modulus E is equivalent to that, in terms of distance of mutual approach, radius of circle of contact, and indenter load, as occurring between a rigid indenter of radius R_i and a specimen with modulus E^* in accordance with Eq. 1.2a. However, physically, the shaded volume of material is not displaced by the indenter and so the contact could also be viewed as occurring between a rigid indenter of radius R^+ and a specimen of modulus E (Courtesy CSIRO).

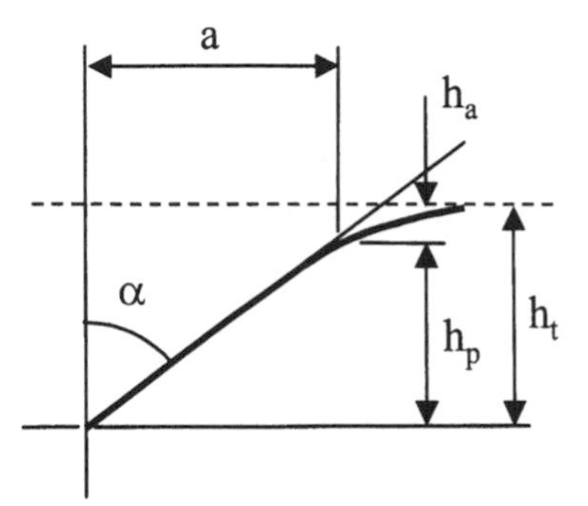

Fig. 1.3 Geometry of contact with conical indenter.

For a conical indenter, similar equations apply where the radius of circle of contact is related to the indenter load by[7]:

$$P = \frac{\pi a}{2} E^* a \cot \alpha \tag{1.2k}$$

The depth profile of the deformed surface within the area of contact is:

$$h = \left(\frac{\pi}{2} - \frac{r}{a}\right) a \cot \alpha \quad r \leq a \tag{1.2l}$$

where α is the cone semi-angle as shown in Fig. 1.3. The quantity a cot α is the depth of penetration h_p measured at the circle of contact. Substituting Eq. 1.2k into 1.2l with r = 0, we obtain:

$$P = \frac{2E^* \tan \alpha}{\pi} h_t^2 \tag{1.2m}$$

where h_t is the depth of penetration of the apex of the indenter beneath the original specimen free surface.

In indentation testing, the most common types of indenters are spherical indenters, where the Hertz equations apply directly, or pyramidal indenters. The most common types of pyramidal indenters are the four-sided Vickers indenter and the three-sided Berkovich indenter. Of particular interest in indentation testing is the area of the contact found from the dimensions of the contact perimeter. For a spherical indenter, the radius of the circle of contact is given by:

$$\begin{aligned} a &= \sqrt{2R_i h_p - h_p^2} \\ &\approx \sqrt{2R_i h_p} \end{aligned} \tag{1.2n}$$

where h_p is the depth of the circle of contact as shown in Fig. 1.1. The approximation of Eq. 1.2n is precisely that which underlies the Hertz equations (Eqs. 1.2a and 1.2d) and thus these equations apply to cases where the deformation is small, that is, when the depth h_p is small in comparison to the radius R_i.

For a conical indenter, the radius of the circle of contact is simply:

$$a = h_p \tan\alpha \tag{1.2o}$$

Table 1.1 Projected areas, intercept corrections, and geometry correction factors for various types of indenters. The semi-angles given for pyramidal indenters are the face angles with the central axis of the indenter.

Indenter type	Projected area	Semi-angle θ (deg)	Effective cone angle α (deg)	Intercept factor* ε	Geometry correction factor β
Sphere	$A \approx \pi 2Rh_p$	N/A	N/A	0.75	1
Berkovich	$A = 3\sqrt{3}h_p^2 \tan^2\theta$	65.27°	70.3°	0.75	1.034
Vickers	$A = 4h_p^2 \tan^2\theta$	68°	70.3°	0.75	1.012
Knoop	$A = 2h_p^2 \tan\theta_1 \tan\theta_2$	$\theta_1 =$ 86.25°, $\theta_2 =$ 65°	77.64°	0.75	1.012
Cube Corner	$A = 3\sqrt{3}h_p^2 \tan^2\theta$	35.26°	42.28°	0.75	1.034
Cone	$A = \pi h_p^2 \tan^2\alpha$	α	α	0.727	1

In indentation testing, pyramidal indenters are generally treated as conical indenters with a cone angle that provides the same area to depth relationship as the actual indenter in question. This allows the use of convenient axial-symmetric elastic equations, Eqs. 1.2k to 1.2m, to be applied to contacts involving non-axial-symmetric indenters. Despite the availability of contact solutions for pyramidal punch problems,[8,9,10] the conversion to an equivalent axial-symmetric has found a wide acceptance.

The areas of contact as a function of the depth of the circle of contact for some common indenter geometries are given in Table 1.1 along with other information to be used in the analysis methods shown in Chapter 3.

* The intercept factors given here are those most commonly used. The values for the pyramidal indenters should theoretically be 0.72 but it has been shown that a value of 0.75 better represents experimental data (see Chapter 3).

1.3 Geometrical Similarity

With a pyramidal or conical indenter, the ratio of the length of the diagonal or radius of circle of contact to the depth of the indentation,† a/δ, remains constant for increasing indenter load, as shown in Fig. 1.4. Indentations of this type have the property of "geometrical similarity." For geometrically similar indentations, it is not possible to set the scale of an indentation without some external reference. The significance of this is that the strain within the material is a constant, independent of the load applied to the indenter.

Unlike a conical indenter, the radius of the circle of contact for a spherical indenter increases faster than the depth of the indentation as the load increases. The ratio a/δ increases with increasing load. In this respect, indentations with a spherical indenter are not geometrically similar. Increasing the load on a spherical indenter is equivalent to decreasing the tip semi-angle of a conical indenter.

However, geometrically similar indentations may be obtained with spherical indenters of different radii. If the indentation strain, a/R, is maintained constant, then so is the mean contact pressure, and the indentations are geometrically similar.

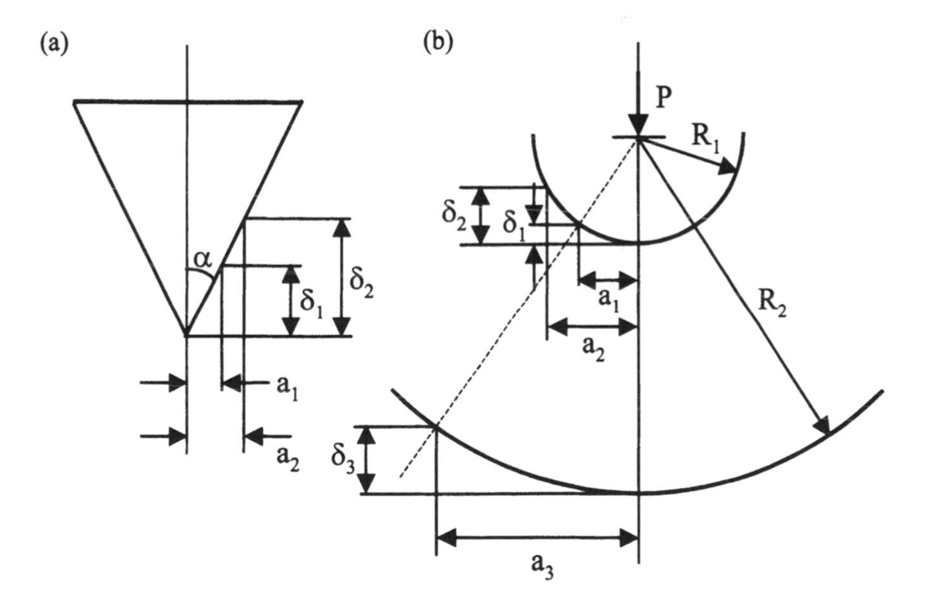

Fig. 1.4 Geometrical similarity for (a) diamond pyramid or conical indenter; (b) spherical indenter. For the conical indenter, $a_1/\delta_1 = a_2/\delta_2$. For the spherical indenter, $a_1/\delta_1 <> a_2/\delta_2$ but $a_1/\delta_1 = a_3/\delta_3$ if $a_1/R_1 = a_3/R_3$ (after reference 11).

† In this section only, δ is the indentation depth measured from the contact circle, not below the original free surface.

The principle of geometrical similarity is widely used in hardness measurements. For example, owing to geometrical similarity, hardness measurements made using a diamond pyramid indenter are expected to yield a value for hardness that is independent of the load. For spherical indenters, the same value of mean contact pressure may be obtained with different sized indenters and different loads as long as the ratio of the radius of the circle of contact to the indenter radius, a/R, is the same in each case.

The quantity a/R for a spherical indentation is equivalent to cot α for a conical indenter. Tabor[12] showed that the representative strain in a Brinell hardness test is equal to about 0.2a/R and hence the representative strain in a typical indentation test performed with a Vickers indenter is approximately 8% (setting $\alpha = 68°$). This is precisely the indentation strain at which a fully developed plastic zone is observed to occur in the Brinell hardness test.

1.4 Elastic–Plastic Contact

Indentation tests on many materials result in both elastic and plastic deformation of the specimen material. In brittle materials, plastic deformation most commonly occurs with pointed indenters such as the Vickers diamond pyramid. In ductile materials, plasticity may be readily induced with a "blunt" indenter such as a sphere or cylindrical punch. Indentation tests are used routinely in the measurement of hardness of materials, but Vickers, Berkovich, and Knoop diamond indenters may be used to investigate other mechanical properties of solids such as specimen strength, fracture toughness, and internal residual stresses. The meaning of hardness has been the subject of considerable attention by scientists and engineers since the early 1700s. It was appreciated very early on that hardness indicated a resistance to penetration or permanent deformation. Early methods of measuring hardness, such as the scratch method, although convenient and simple, were found to involve too many variables to provide the means for a scientific definition of hardness. Static indentation tests involving spherical or conical indenters were first used as the basis for theories of hardness. Compared to "dynamic" tests, static tests enabled various criteria of hardness to be established since the number of test variables was reduced to a manageable level. The most well-known criterion is that of Hertz, who postulated that an absolute value for hardness was the least value of pressure beneath a spherical indenter necessary to produce a permanent set at the center of the area of contact. Later treatments by Auerbach,[13] Meyer,[14] and Hoyt[15] were all directed to removing some of the practical difficulties in Hertz's original proposal.

1.4.1 The Constraint Factor

Static indentation hardness tests usually involve the application of load to a spherical or pyramidal indenter. The pressure distribution beneath the indenter is of particular interest. The value of the mean contact pressure p_m at which there is no increase with increasing indenter load is shown by experiment to be related to the hardness number H. For hardness methods that employ the projected contact area, the hardness number H is given directly by the mean pressure p_m at this limiting condition. Experiments show that the mean pressure between the indenter and the specimen is directly proportional to the material's yield, or flow stress in compression, and can be expressed as:

$$H \approx CY \tag{1.4.1a}$$

where Y is the yield, or flow stress, of the material. The mean contact pressure in an indentation test is higher than that required to initiate yield in a uniaxial compression test because it is the shear component of stress that is responsible for plastic flow. The maximum shear stress is equal to half the difference between the maximum and minimum principal stresses, and in an indentation stress field, where the stress material is constrained by the surrounding matrix, there is a considerable hydrostatic component. Thus, the mean contact pressure is greater than that required to initiate yield when compared to a uniaxial compressive stress. It is for this reason that C in Eq. 1.4.1a is called the "constraint factor," the value of which depends upon the type of specimen, the type of indenter, and other experimental parameters. For the indentation methods mentioned here, both experiments and theory predict $C \approx 3$ for materials with a large value of the ratio E/Y (e.g., metals). For low values of E/Y (e.g., glasses[16,17]), $C \approx 1.5$. The flow, or yield stress Y, in this context is the stress at which plastic yielding first occurs.

1.4.2 Indentation Response of Materials

A material's hardness value is intimately related to the mean contact pressure p_m beneath the indenter at a limiting condition of compression. Valuable information about the elastic and plastic properties of a material can be obtained with spherical indenters when the mean contact pressure, or "indentation stress," is plotted against the ratio a/R, the "indentation strain". The indentation stress–strain response of an elastic–plastic solid can generally be divided into three regimes, which depend on the uniaxial compressive yield stress Y of the material[12]:

1. $p_m < 1.1Y$ — full elastic response, no permanent or residual impression left in the test specimen after removal of load.
2. $1.1Y < p_m < CY$ — plastic deformation exists beneath the surface but is constrained by the surrounding elastic material, where C is a constant whose value depends on the material and the indenter geometry.

3. $p_m = CY$ — plastic region extends to the surface of the specimen and continues to grow in size such that the indentation contact area increases at a rate that gives little or no increase in the mean contact pressure for further increases in indenter load.

In Region 1, during the initial application of load, the response is elastic and can be predicted from Eq. 1.2j. Equation 1.2j assumes linear elasticity and makes no allowance for yield within the specimen material. For a fully elastic response, the principal shear stress for indentation with a spherical indenter is a maximum at $\approx 0.47p_m$ at a depth of $\approx 0.5a$ beneath the specimen surface directly beneath the indenter.[18] Following Tabor,[12] either the Tresca or von Mises shear stress criteria may be employed, where plastic flow occurs at $\tau \approx 0.5Y$, to show that plastic deformation in the specimen beneath a spherical indenter can be first expected to occur when $p_m \approx 1.1Y$. Theoretical treatment of events within Region 2 is difficult because of the uncertainty regarding the size and shape of the evolving plastic zone. At high values of indentation strain (Region 3), the mode of deformation appears to depend on the type of indenter and the specimen material. The presence of the free surface has an appreciable effect, and the plastic deformation within the specimen is such that, assuming no work hardening, little or no increase in p_m occurs with increasing indenter load.

1.4.3 Elastic–Plastic Stress Distribution

The equations for elastic contact given above form the basis of analysis methods for nanoindentation tests, even if these tests involve plastic deformation in the specimen. Hertz's original analysis was concerned with the form of the pressure distribution between contacting spheres that took the form:

$$\frac{\sigma_z}{p_m} = -\frac{3}{2}\left(1 - \frac{r^2}{a^2}\right)^{1/2} \tag{1.4.3a}$$

The pressure distribution σ_z in Eq. 1.4.3a is normalized to the mean contact pressure p_m, and it can be seen that the pressure is a maximum equal to 1.5 times the mean contact pressure at the center of the contact as shown in Fig. 1.5 (c).[19]

When plastic deformation occurs, the pressure distribution is modified and becomes more uniform. Finite element results for an elastic–plastic contact are shown in Fig. 1.5 (c). There is no currently available analytical theory that generally describes the stress distribution beneath the indenter for an elastic–plastic contact.

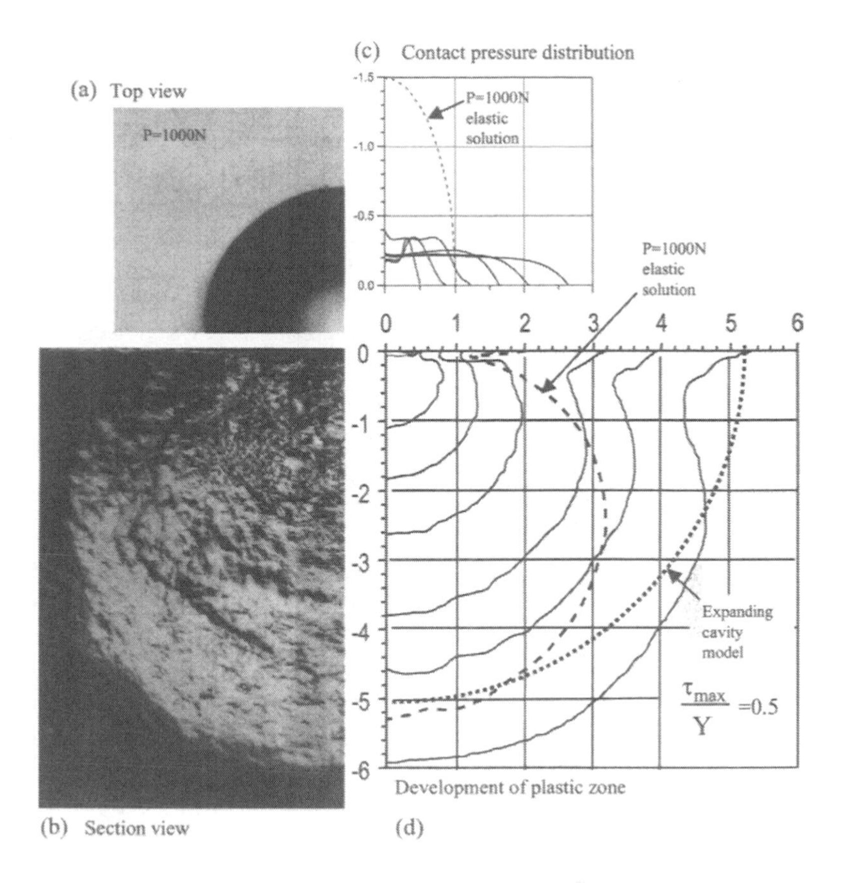

Fig. 1.5 Elastic–plastic indentation response for mild steel material, E/Y = 550. (a) Test results for an indenter load of P = 1000 N and indenter of radius 3.18 mm showing residual impression in the surface. (b) Section view with subsurface accumulated damage beneath the indentation site. (c) Finite element results for contact pressure distribution. (d) Finite element results showing development of the plastic zone in terms of contours of maximum shear stress at $\tau_{max}/Y = 0.5$. In (c) and (d), results are shown for indentation strains of a/R = 0.04, 0.06, 0.08, 0.11, 0.14, 0.18. Distances are expressed in terms of the contact radius a = 0.218 mm for the elastic case of P = 1000 N (after reference 19).

The finite element method, however, has been used with some success in this regard and Fig. 1.5 (d) shows the evolution of the plastic zone (within which the shear stress is a constant) compared with an experiment on a specimen of mild steel, Fig. 1.5 (b). Mesarovic and Fleck[20] have calculated the full elastic elastic–plastic contact for a spherical indenter that includes elasticity, strain-hardening, and interfacial friction.

1.4.4 Hardness Theories

Theoretical approaches to hardness can generally be categorized according to the characteristics of the indenter and the response of the specimen material. Various semi-empirical models that describe experimentally observed phenomena at values of indentation strain at or near a condition of a fully developed plastic zone have been given considerable attention in the literature.[12,17,21–33] These models variously describe the response of the specimen material in terms of slip lines, elastic displacements, and radial compressions. For sharp wedge or conical indenters, substantial upward flow is usually observed, and because elastic strains are thus negligible compared to plastic strains, the specimen can be regarded as being rigid–plastic. A cutting mechanism is involved, and new surfaces are formed beneath the indenter as the volume displaced by the indenter is accommodated by the upward flow of plastically deformed material. The constraint factor C in this case arises due to flow and velocity considerations.[21] For blunt indenters, the specimen responds in an elastic–plastic manner, and plastic flow is usually described in terms of the elastic constraint offered by the surrounding material. With blunt indenters, Samuels and Mulhearn[24] noted that the mode of plastic deformation at a condition of fully developed plastic zone appears to be a result of compression rather than cutting, and the displaced volume of the indenter is assumed to be taken up entirely by elastic strains within the specimen material outside the plastic zone. This idea was given further attention by Marsh,[23] who compared the plastic deformation in the vicinity of the indenter to that which occurs during the radial expansion of a spherical cavity subjected to internal pressure, an analysis of which was given previously by Hill.[22] The most widely accepted treatment is that of Johnson,[26,33] who replaced the expansion of the cavity with that of an incompressible hemispherical core of material subjected to an internal pressure. Here, the core pressure is directly related to the mean contact pressure. This is the so-called "expanding cavity" model.

In the expanding cavity model, the contacting surface of the indenter is encased by a hydrostatic "core" of radius a_c, which is in turn surrounded by a hemispherical plastic zone of radius c as shown in Fig. 1.6. An increment of penetration dh of the indenter results in an expansion of the core da and the volume displaced by the indenter is accommodated by radial movement of particles du(r) at the core boundary. This in turn causes the plastic zone to increase in radius by an amount dc.

For geometrically similar indentations, such as with a conical indenter, the radius of the plastic zone increases at the same rate as that of the core, hence, $da/dc = a/c$.

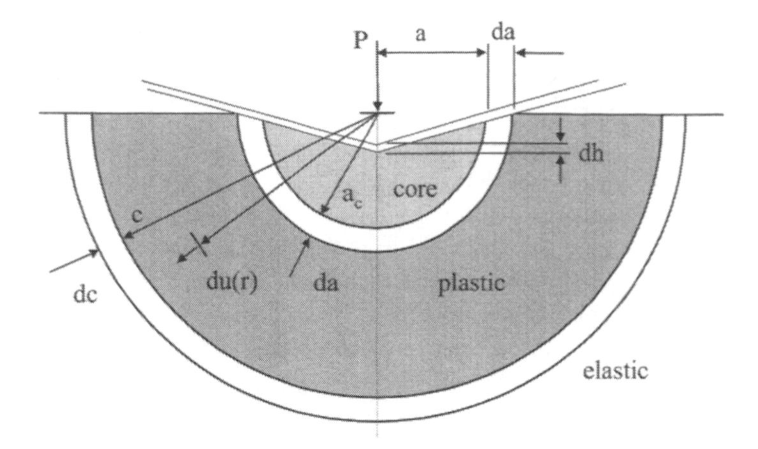

Fig. 1.6 Expanding cavity model schematic. The contacting surface of the indenter is encased by a hydrostatic "core" of radius a_c that is in turn surrounded by a hemispherical plastic zone of radius c. An increment of penetration dh of the indenter, results in an expansion of the core da and the volume displaced by the indenter is accommodated by radial movement of particles du(r) at the core boundary. This in turn causes the plastic zone to increase in radius by an amount dc (after reference 19).

Using this result Johnson shows that the pressure in the core can be calculated from:

$$\frac{p}{Y}=\frac{2}{3}\left[1+\ln\left(\frac{(E/Y)\tan\beta+4(1-2\nu)}{6\left(1-\nu^2\right)}\right)\right] \tag{1.4.4a}$$

where p is the pressure within the core and β is the angle of inclination of the indenter with the specimen surface.

The mean contact pressure is found from:

$$p_m = p+\frac{2}{3}Y \tag{1.4.4b}$$

and this leads to an value for the constraint factor C. When the free surface of the specimen begins to influence appreciably the shape of the plastic zone, and the plastic material is no longer elastically constrained, the volume of material displaced by the indenter is accommodated by upward flow around the indenter. The specimen then takes on the characteristics of a rigid–plastic solid, because any elastic strains present are very much smaller than the plastic flow of unconstrained material. Plastic yield within such a material depends upon a critical shear stress, which may be calculated using either of the von Mises or Tresca failure criteria.

In the slip-line field solution, developed originally in two dimensions by Hill, Lee, and Tupper,[21] the volume of material displaced by the indenter is accounted for by upward flow, as shown in Fig. 1.7. The material in the region ABCDE flows upward and outward as the indenter moves downward under load. Because frictionless contact is assumed, the direction of stress along the line AB is normal to the face of the indenter. The lines within the region ABDEC are oriented at 45° to AB and are called "slip lines" (lines of maximum shear stress). This type of indentation involves a "cutting" of the specimen material along the line 0A and the creation of new surfaces that travel upward along the contact surface. The contact pressure across the face of the indenter is:

$$\begin{aligned} p_m &= 2\tau_{max}(1+\alpha) \\ &= H \end{aligned} \tag{1.4.4c}$$

where τ_{max} is the maximum value of shear stress in the specimen material and α is the cone semi-angle (in radians).

Invoking the Tresca shear stress criterion, where plastic flow occurs at τ_{max} = 0.5Y, and substituting into Eq. 1.4.4c, gives:

$$\begin{aligned} H &= Y(1+\alpha) \\ &\therefore \\ C &= (1+\alpha) \end{aligned} \tag{1.4.4d}$$

The constraint factor determined by this method is referred to as C_{flow}. For values of α between 70° and 90°, Eq. 1.4.4b gives only a small variation in C_{flow} of 2.22 to 2.6. Friction between the indenter and the specimen increases the value of C_{flow}. A slightly larger value for C_{flow} is found when the von Mises stress criterion is used (where $\tau_{max} \approx 0.58Y$). For example, at $\alpha = 90°$, Eq. 1.3.4b with the von Mises criterion gives C = 3.

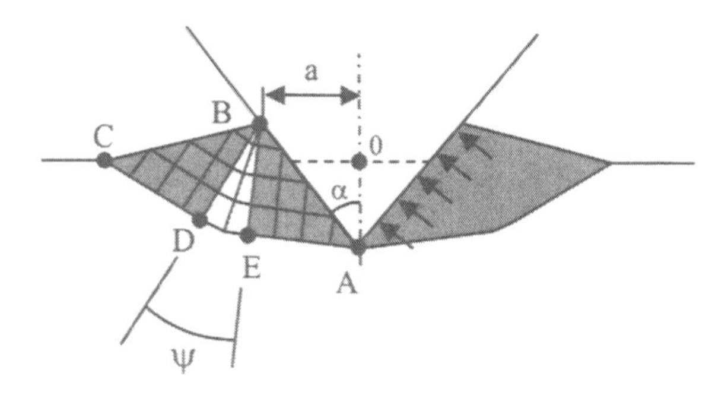

Fig. 1.7 Slip-line theory. The material in the region ABCDE flows upward and outward as the indenter moves downward under load (after reference 11).

1.5 Indentations at the Nanometre Scale

The present field of nanoindentation grew from a desire to measure the mechanical properties of hard thin films and other near surface treatments in the early 1980s. Microhardness testing instruments available at the time could not apply low enough forces to give penetration depths less than the required 10% or so of the film thickness so as to avoid influence on the hardness measurement from the presence of the substrate. Even if they could, the resulting size of the residual impression cannot be determined with sufficient accuracy to be useful. For example, the uncertainty in a measurement of a 5 μm diagonal of a residual impression made by a Vickers indenter is on the order of 20% when using an optical method and increases with decreasing size of indentation and can be as high as 100% for a 1 μm impression.

Since the spatial dimensions of the contact area are not conveniently measured, modern nanoindentation techniques typically use the measured depth of penetration of the indenter and the known geometry of the indenter to determine the contact area. Such a procedure is sometimes called "depth-sensing indentation testing" although it is quite permissible to use the technique at macroscopic dimensions.[34,35] For such a measurement to be made, the depth measurement system needs to be referenced to the specimen surface, and this is usually done by bringing the indenter into contact with the surface with a very small "initial contact force," which, in turn, results in an inevitable initial penetration of the surface by the indenter that must be accounted for in the analysis. Additional corrections are required to account for irregularities in the shape of the indenter, deflection of the loading frame, and piling-up of material around the indenter (see Fig. 1.8). These effects contribute to errors in the recorded depths and, subsequently, the hardness and modulus determinations. Furthermore, the scale of deformation in a nanoindentation test becomes comparable to the size of material defects such as dislocations and grain sizes, and the continuum approximation used in the analysis can become less valid.

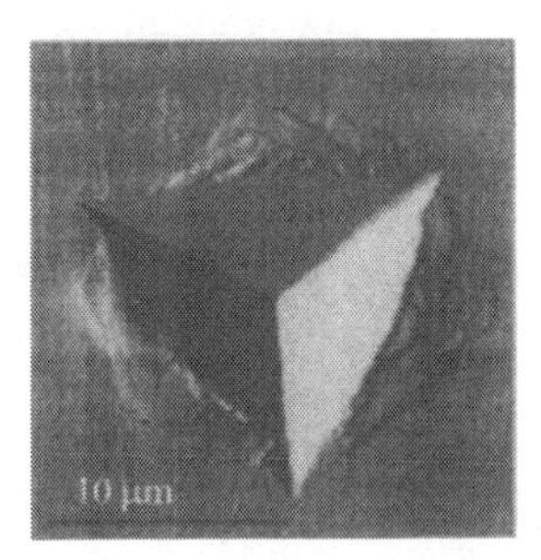

Fig. 1.8. Atomic force micrograph of a residual impression in steel made with a triangular pyramid Berkovich indenter. Note the presence of piling-up at the periphery of the contact impression (Courtesy CSIRO).

The nanoindentation test results provide information on the elastic modulus, hardness, strain-hardening, cracking, phase transformations, creep, and energy absorption. The specimen size is very small and the test can in many cases be considered non-destructive. Specimen preparation is straightforward. Because the scale of deformation is very small, the technique is applicable to thin surface films and surface modified layers. In many cases, the microstructural features of a thin film or coating differs markedly from that of the bulk material owing to the presence of residual stresses, preferred orientations of crystallographic planes, and the morphology of the microstructure. The applications of the technique therefore cover technologies such as cathodic arc deposition, physical vapor deposition (PVD), and chemical vapor deposition (CVD) as well as ion-implantation and functionally graded materials. Nanoindentation instruments are typically easy to use, operate under computer control, and require no vacuum chambers or other expensive laboratory infrastructure.

The technique relies on a continuous measurement of depth of penetration with increasing load. Such measurements at the micron scale were demonstrated by Fröhlich, Grau and Grellmann[36] in 1977 who analyzed the loading and unloading curves for a variety of materials and foreshadowed the use of the technique for the measurement of surface properties of materials. Pethica,[37] in 1981, applied the method to the measurement of the mechanical properties of ion-implanted metal surfaces, a popular application of the technique for many years.[38] Stilwell and Tabor[39] focused on the elastic unloading of the impression as did Armstrong and Robinson[40] in 1974, and Lawn and Howes[41] in 1981. The present modern treatments probably begin with Bulychev, Alekhin, Shorshorov, and Ternovskii[42], who in 1975 showed how the area of contact could be measured using the unloading portion of the load-displacement curve. Loubet, Georges, Marchesini, and Meille[43] used this method for relatively high load testing (in the order of 1 Newton) and Doerner and Nix[44] extended the measurements into the millinewton range in 1986. The most commonly used method of analysis is a refinement of the Doerner and Nix approach by Oliver and Pharr[45] in 1992. A complementary approach directed to indentations with spherical indenters was proposed by Field and Swain and coworkers[46,47] in 1993 and subsequently shown to be equivalent to the Oliver and Pharr method.[48] Review articles[49–51] on micro, and nanoindentation show a clear evolution of the field from traditional macroscopic measurements of hardness. The field now supports specialized symposia on an annual basis attracting papers covering topics from fundamental theory to applications of the technique.

The first "ultra-micro" hardness tests were done with apparatus designed for use inside the vacuum chamber of a scanning electron microscope (SEM), where load was applied to a sharply pointed tungsten wire via the movement of a galvanometer that was controlled externally by electric current. Depth of penetration was determined by measuring the motion of the indenter support using an interferometric method. The later use of strain gauges to measure the applied load and finely machined parallel springs operated by an electromagnetic coil bought the measurement outside the vacuum chamber into the laboratory, but, although the required forces could now be applied in a controlled manner, optical measurements of displacement or sizes of residual impressions remained a limiting factor. Developments in electronics lead to the production of displacement measuring sensors with resolutions greater than those offered by optical methods and, in the last ten years, some six or seven instruments have evolved into commercial products, often resulting in the creation of private companies growing out of research organizations to sell and support them.

A common question is to ask if it is possible to perform nanoindentation with an atomic force microscope (AFM). It is, but there are some significant problems that make the method unsuitable for precise determination of materials properties. In an AFM, the tip is usually grown from silicon with a tip radius in the order of 5 – 10 nm. The precise geometry is not usually known, since for imaging purposes, it is not necessary to know this information. The cantilever, to which the tip is mounted, is a very compliant structure – made so since the purpose of the instrument is to provide a large deflection in response to surface forces. Thus, unlike a conventional nanoindentation instrument, a large component of a load-displacement curve obtained with an AFM is accounted for by compliance of the cantilever rather than indentation into the specimen, the latter being the signal of interest in nanoindentation studies. Any instrumented technique (such as an AFM or nanoindentation instrument) employs analog to digital conversion of data and since the signal of interest (the penetration depth) from an indentation made with an AFM is such a small proportion of the total signal, there is a considerable loss of resolution in this quantity in the overall results. This, together with the uncertainties regarding the tip shape, limit the use of this type of instrument for extraction of material properties from the indentation data.

There is no doubt that as the scale of mechanisms becomes smaller, interest in mechanical properties on a nanometre scale and smaller, and the nature of surface forces and adhesion, will continue to increase. Indeed, at least one recent publication refers to the combination of a nanoindenter and an atomic force microscope as a "picoindenter"[52] suitable for the study of pre-contact mechanics, the process of making contact, and actual contact mechanics. The present maturity of the field of nanoindentation makes it a suitable technique for the evaluation of new materials technologies by both academic and private industry research laboratories and is increasingly finding application as a quality control tool.

References

1. H. Hertz, "On the contact of elastic solids," J. Reine Angew. Math. 92, 1881, pp. 156–171. Translated and reprinted in English in *Hertz's Miscellaneous Papers*, Macmillan & Co., London, 1896, Ch. 5.
2. H. Hertz, "On hardness," Verh. Ver. Beförderung Gewerbe Fleisses 61, 1882, p. 410. Translated and reprinted in English in *Hertz's Miscellaneous Papers*, Macmillan & Co, London, 1896, Ch. 6.
3. S. Timoshenko and J.N. Goodier, *Theory of Elasticity*, 2nd Ed. McGraw-Hill, N.Y. 1951.
4. A.C. Fischer-Cripps, "The use of combined elastic modulus in the analysis of depth sensing indentation data," J. Mater. Res. 16 11, 2001, pp. 3050–3052.
5. M.M. Chaudhri, "A note on a common mistake in the analysis of nanoindentation test data", J. Mater. Res. 16 2, 2001, pp. 336–339.
6. A.C. Fischer-Cripps, "The use of combined elastic modulus in depth-sensing indentation with a conical indenter," J. Mat. Res. 18 5, 2003, pp.1043–1045..
7. I.N. Sneddon, "Boussinesq's problem for a rigid cone," Proc. Cambridge Philos. Soc. 44, 1948, pp. 492-507.
8. J.R. Barber and D.A. Billings, "An approximate solution for the contact area and elastic compliance of a smooth punch of arbitrary shape," Int. J. Mech. Sci. 32 12, 1990, pp. 991–997.
9. G.G. Bilodeau, "Regular pyramid punch problem," J. App. Mech. 59, 1992, pp. 519–523.
10. P.-L. Larsson, A.E. Giannakopolous, E. Soderlund, D.J. Rowcliffe and R. Vestergaard, "Analysis of Berkovich Indentation," Int. J. Structures, 33 2, 1996, pp.221–248.
11. A.C. Fischer-Cripps, *Introduction to Contact Mechanics*, Springer-Verlag, New York, 2000.
12. D. Tabor, *The Hardness of Metals*, Clarendon Press, Oxford, 1951.
13. F. Auerbach, "Absolute hardness," Ann. Phys. Chem. (Leipzig) 43, 1891, pp.61–100. Translated by C. Barus, Annual Report of the Board of Regents of the Smithsonian Institution, July 1, 1890 – June 30 1891, reproduced in "Miscellaneous documents of the House of Representatives for the First Session of the Fifty-Second Congress," Government Printing Office, Washington, D.C., 43, 1891–1892, pp.207–236.
14. E. Meyer, "Untersuchungen über Harteprufung und Harte," Phys. Z. 9, 1908, pp. 66–74.
15. S.L. Hoyt, "The ball indentation hardness test," Trans. Am. Soc. Steel Treat. 6, 1924, pp. 396–420.
16. M.C. Shaw, "The fundamental basis of the hardness test," in *The Science of Hardness Testing and its Research Applications*, J.H. Westbrook and H. Conrad, Eds. American Society for Metals, Cleveland, OH, 1973, pp. 1–15.
17. M.V. Swain and J.T. Hagan, "Indentation plasticity and the ensuing fracture of glass," J. Phys. D: Appl. Phys. 9, 1976, pp. 2201–2214.

18. M.T. Huber, "Contact of solid elastic bodies," Ann. D. Physik, 14 1, 1904, pp. 153–163.
19. A.C. Fischer-Cripps, "Elastic–plastic response of materials loaded with a spherical indenter," J. Mater. Sci. 32 3, 1997, pp. 727–736.
20. S.Dj. Mesarovic and N. A. Fleck, "Spherical indentation of elastic–plastic solids," Proc. Roy. Soc. A455, 1999, pp. 2707–2728.
21. R. Hill, E.H. Lee and S.J. Tupper, "Theory of wedge-indentation of ductile metals," Proc. Roy. Soc. A188, 1947, pp. 273–289.
22. R. Hill, *The Mathematical Theory of Plasticity*, Clarendon Press, Oxford, 1950.
23. D.M. Marsh, "Plastic flow in glass," Proc. Roy. Soc. A279, 1964, pp. 420–435.
24. L.E. Samuels and T.O. Mulhearn, "An experimental investigation of the deformed zone associated with indentation hardness impressions," J. Mech. Phys. Solids, 5, 1957, pp. 125–134.
25. T.O. Mulhearn, "The deformation of metals by Vickers-type pyramidal indenters," J. Mech. Phys. Solids, 7, 1959, pp. 85–96.
26. K.L. Johnson, "The correlation of indentation experiments," J. Mech. Phys. Sol. 18, 1970, pp. 115–126.
27. M.C. Shaw and D.J. DeSalvo, "A new approach to plasticity and its application to blunt two dimension indenters," J. Eng. Ind. Trans. ASME, 92, 1970, pp. 469–479.
28. M.C. Shaw and D.J. DeSalvo, "On the plastic flow beneath a blunt axisymmetric indenter," J. Eng. Ind. Trans. ASME 92, 1970, pp. 480–494.
29. C. Hardy, C.N. Baronet, and G.V. Tordion, "The elastic–plastic indentation of a half-space by a rigid sphere," Int. J. Numer. Methods Eng. 3, 1971, pp. 451–462.
30. C.M. Perrott, "Elastic–plastic indentation: Hardness and fracture," Wear 45, 1977, pp. 293–309.
31. S.S. Chiang, D.B. Marshall, and A.G. Evans, "The response of solids to elastic/plastic indentation. 1. Stresses and residual stresses," J. Appl. Phys. 53 1, 1982, pp. 298–311.
32. S.S. Chiang, D.B. Marshall, and A.G. Evans, "The response of solids to elastic/plastic indentation. 2. Fracture initiation," J. Appl. Phys. 53 1, 1982, pp. 312–317.
33. K.L. Johnson, *Contact Mechanics*, Cambridge University Press, Cambridge, 1985.
34. J.H. Ahn and D. Kwon, "Derivation of plastic stress-strain relationship from ball indentations: Examination of strain definition and pileup effect," J. Mater. Res. 16 11, 2001, pp. 3170–3178.
35. J. Thurn, D.J. Morris, and R.F. Cook, "Depth-sensing indentation at macroscopic dimensions," J. Mater. Res. 17 10, 2002, pp. 2679–2690.
36. F. Frölich, P. Grau, and W. Grellmann, "Performance and analysis of recording microhardness tests," Phys. Stat. Sol. (a), 42, 1977, pp. 79–89.
37. J.B. Pethica, "Microhardness tests with penetration depths less than ion implanted layer thickness in ion implantation into metals," Third International Conference on Modification of Surface Properties of Metals by Ion-Implantation, Manchester, England, 23-26, 1981, V. Ashworth et al. eds., Pergammon Press, Oxford, 1982, pp. 147–157.
38. J.S. Field, "Understanding the penetration resistance of modified surface layers," Surface and Coatings Technology, 36, 1988, pp. 817–827.

39. N.A. Stilwell and D. Tabor, "Elastic recovery of conical indentations," Phys. Proc. Soc. 78 2, 1961, pp. 169–179.
40. R.W. Armstrong and W.H. Robinson, "Combined elastic and plastic deformation behaviour from a continuous indentation hardness test," New Zealand Journal of Science, 17, 1974, pp. 429–433.
41. B.R. Lawn and V.R. Howes, "Elastic recovery at hardness indentations," J. Mat. Sci. 16, 1981, pp. 2745–2752.
42. S.I. Bulychev, V.P. Alekhin, M. Kh. Shorshorov, and A.P. Ternorskii, "Determining Young's modulus from the indenter penetration diagram," Zavod. Lab. 41 9, 1975, pp. 11137–11140.
43. J.L. Loubet, J.M. Georges, O. Marchesini, and G. Meille, "Vicker's indentation of magnesium oxide," J. Tribol. 106, 1984, pp. 43–48.
44. M.F. Doerner and W.D. Nix, "A method for interpreting the data from depth-sensing indentation instruments," J. Mater. Res. 1 4, 1986, pp. 601–609.
45. W.C. Oliver and G.M. Pharr, "An improved technique for determining hardness and elastic modulus using load and displacement sensing indentation experiments," J. Mater. Res. 7 4, 1992, pp. 1564–1583.
46. T.J. Bell, A. Bendeli, J.S. Field, M.V. Swain, and E.G. Thwaite, "The determination of surface plastic and elastic properties by ultra-micro indentation," Metrologia, 28, 1991, pp. 463–469.
47. J.S. Field and M.V. Swain, "A simple predictive model for spherical indentation," J. Mater. Res. 8 2, 1993, pp. 297–306.
48. A.C. Fischer-Cripps, "Study of analysis methods for depth-sensing indentation test data for spherical indenters," J. Mater. Res. 16 6, 2001, pp. 1579–1584.
49. A.G. Atkins, "Topics in indentation hardness," Metal Science, 16, 1982, pp. 127–137.
50. H.M. Pollock, "Nanoindentation", ASM Handbook, Friction, Lubrication, and Wear Technology, 18, 1992, pp. 419–429.
51. J.L. Hay and G.M. Pharr, "Instrumented indentation testing," ASM Handbook, Materials Testing and Evaluation, 8, 2000, pp. 232–243.
52. S.A. Syed, K.J. Wahl, and R.J. Colton, "Quantitative study of nanoscale contact and pre-contact mechanics using force modulation," Mat. Res. Soc. Symp. Proc. 594, 2000, pp. 471–476.

Chapter 2
Nanoindentation Testing

2.1 Nanoindentation Test Data

The goal of the majority of nanoindentation tests is to extract elastic modulus and hardness of the specimen material from load-displacement measurements. Conventional indentation hardness tests involve the measurement of the size of a residual plastic impression in the specimen as a function of the indenter load. This provides a measure of the area of contact for a given indenter load. In a nanoindentation test, the size of the residual impression is often only a few microns and this makes it very difficult to obtain a direct measure using optical techniques. In nanoindentation testing, the depth of penetration beneath the specimen surface is measured as the load is applied to the indenter. The known geometry of the indenter then allows the size of the area of contact to be determined. The procedure also allows for the modulus of the specimen material to be obtained from a measurement of the "stiffness" of the contact, that is, the rate of change of load and depth. In this chapter, the mechanics of the actual indentation test and the nature of the indenters used in this type of testing are reviewed.

2.2 Indenter Types

Nanoindentation hardness tests are generally made with either spherical or pyramidal indenters. Consider a Vickers indenter with opposing faces at a semi-angle of $\theta = 68°$ and therefore making an angle $\beta = 22°$ with the flat specimen surface. For a particular contact radius a, the radius R of a spherical indenter whose edges are at a tangent to the point of contact with the specimen is given by $\sin \beta = a/R$, which for $\beta = 22°$ gives $a/R = 0.375$. It is interesting to note that this is precisely the indentation strain‡ at which Brinell hardness tests, using a spherical indenter, are generally performed, and the angle $\theta = 68°$ for the Vickers indenter was chosen for this reason.

‡ Recall that the term "indentation strain" refers to the ratio a/R.

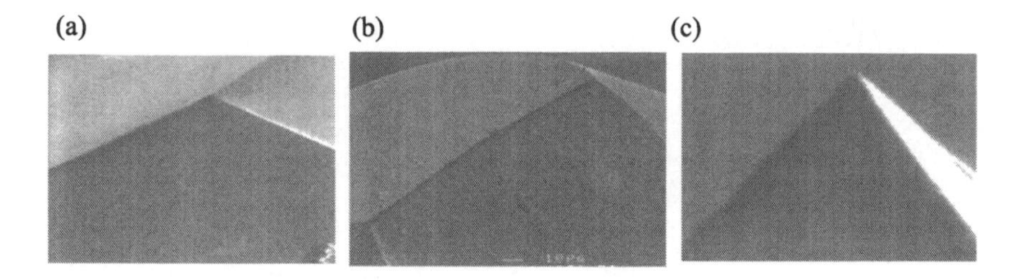

Fig. 2.1 SEM images of the tips of (a) Berkovich, (b) Knoop, and (c) cube-corner indenters used for nanoindentation testing. The tip radius of a typical diamond pyramidal indenter is in the order of 100 nm (Courtesy CSIRO).

The Berkovich indenter,[1] (a) in Fig. 2.1, is generally used in small-scale indentation studies and has the advantage that the edges of the pyramid are more easily constructed to meet at a single point, rather than the inevitable line that occurs in the four-sided Vickers pyramid. The face angle of the Berkovich indenter normally used for nanoindentation testing is 65.27°, which gives the same projected area-to-depth ratio as the Vickers indenter. Originally, the Berkovich indenter was constructed with a face angle of 65.03°, which gives the same *actual* surface area to depth ratio as a Vickers indenter. The tip radius for a typical new Berkovich indenter is on the order of 50–100nm. This usually increases to about 200 nm with use. The Knoop indenter, (b) in Fig. 2.1, is a four-sided pyramidal indenter with two different face angles. Measurement of the unequal lengths of the diagonals of the residual impression is very useful for investigating anisotropy of the surface of the specimen. The indenter was originally developed to allow the testing of very hard materials where a longer diagonal line could be more easily measured for shallower depths of residual impression. The cube corner indenter, (c) in Fig. 2.1, is finding increasing popularity in nanoindentation testing. It is similar to the Berkovich indenter but has a semi-angle at the faces of 35.26°.

Conical indenters have the advantage of possessing axial symmetry, and, with reference to Fig. 2.1, equivalent projected areas of contact between conical and pyramidal indenters are obtained when:

$$A = \pi h_p^{\,2} \tan^2 \alpha \tag{2.2a}$$

where h_p is depth of penetration measured from the edge of the circle or area of contact. For a Vickers or Berkovich indenter, the projected area of contact is $A = 24.5h^2$ and thus the semi-angle for an equivalent conical indenter is 70.3°. It is convenient when analyzing nanoindentation test data taken with pyramidal indenters to treat the indentation as involving an axial-symmetric conical indenter with an apex semi-angle that can be determined from Eq. 2.2a. Table 1.1 gives expressions for the contact area for different types of pyramidal indenters in terms of the penetration depth h_p for the geometries shown in Fig. 2.2.

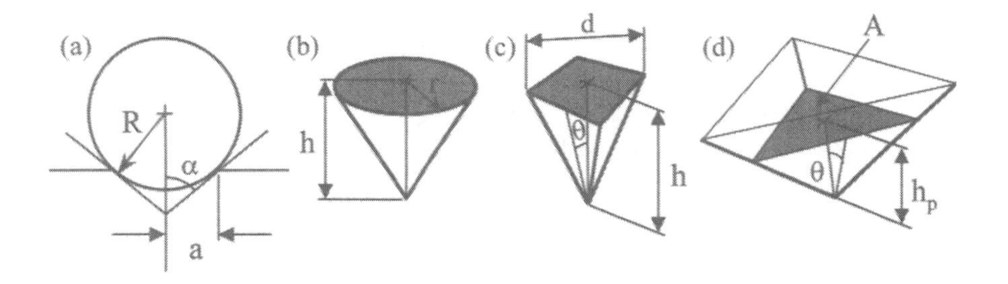

Fig. 2.2 Indentation parameters for (a) spherical, (b) conical, (c) Vickers, and (d) Berkovich indenters (not to scale).

Spherical indenters are finding increasing popularity, as this type of indenter provides a smooth transition from elastic to elastic–plastic contact. It is particularly suitable for measuring soft materials and for replicating contact damage in in-service conditions. As shown in Fig. 2.3, the indenter is typically made as a sphero-cone for ease of mounting. Only the very tip of the indenter is used to penetrate the specimen surface in indentation testing. Diamond spherical indenters with a radius of less than 1 micron can be routinely fashioned.

Indenters can generally be classified into two categories — sharp or blunt. The criteria upon which a particular indenter is classified, however, are the subject of opinion.

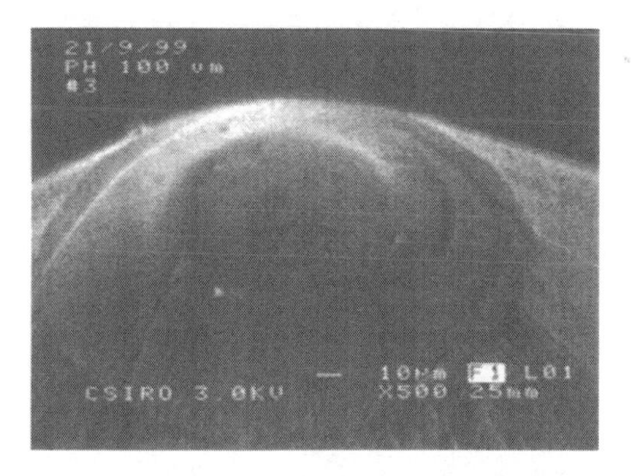

Fig. 2.3 Tip of a sphero-conical indenter used for nanoindentation and scratch testing. Nominal tip radius is 100 μm in this example. Tip radii of <1 μm are available (Courtesy CSIRO).

For example, some authors[2] classify sharp indenters as those resulting in permanent deformation in the specimen upon the removal of load. A Vickers diamond pyramid is such an example in this scheme. However, others prefer to classify a conical or pyramidal indenter with a cone semi-angle $\alpha > 70°$ as being blunt. Thus, a Vickers diamond pyramid with $\theta = 68°$ would in this case be considered blunt. A spherical indenter may be classified as sharp or blunt depending on the applied load according to the angle of the tangent at the point of contact. The latter classification is based upon the response of the specimen material in which it is observed that plastic flow according to the slip-line theory occurs for sharp indenters and the specimen behaves as a rigid–plastic solid. For blunt indenters, the response of the specimen material follows that predicted by the expanding cavity model or the elastic constraint model, depending on the type of specimen material and magnitude of the load. Generally speaking, spherical indenters are termed blunt, and cones and pyramids are sharp.

2.3 Indentation Hardness and Modulus

A particularly meaningful quantity in indentation hardness in the mean contact pressure of the contact, and is found by dividing the indenter load by the projected area of the contact. The mean contact pressure, when determined under conditions of a fully developed plastic zone, is usually defined as the "indentation hardness" H_{IT} of the specimen material. In nanoindentation testing, the displacement of the indenter is measured and the size of the contact area (at full load) is estimated from the depth of penetration with the known geometry of the indenter. For an extreme case of a rigid-plastic solid, where there is little elastic recovery of material, the mean contact pressure at a condition of a fully developed plastic zone is a true representation of the resistance of the material to permanent deformation. When there is substantial elastic recovery, such as in ceramics where the ratio of E/H is high, the mean contact pressure, at a condition of a fully developed plastic zone, is not a true measure of the resistance of the material to plastic deformation but rather measures the resistance of the material to combined elastic and plastic deformations. The distinction is perhaps illustrated by a specimen of rubber, which might deform elastically in an indentation test but undergo very little actual permanent deformation. In this case, the limiting value of mean contact pressure (the apparent indentation hardness) may be very low but the material is actually very resistant to permanent deformation and so the true hardness is very high. The distinction between the true hardness and the apparent hardness is described in more detail in Chapter 3.

In depth-sensing indentation techniques used in nanoindentation, the elastic modulus of the specimen can be determined from the slope of the unloading of the load-displacement response. The modulus measured in this way is formally called the "indentation modulus" E_{IT} of the specimen material. Ideally, the indentation modulus has precisely the same meaning as the term "elastic modulus" or "Young's modulus" but this is not the case for some materials. The value of indentation modulus may be affected greatly by material behavior (e.g. piling-up) that is not accounted for in the analysis of load-displacement data. For this reason, care has to be taken when comparing the modulus for materials generated by different testing techniques and on different types of specimens.

2.3.1 Spherical Indenter

The mean contact pressure, and, hence, indentation hardness, for an impression made with a spherical indenter is given by:

$$p_m = H = \frac{P}{A} = \frac{4P}{\pi d^2} \tag{2.3.1a}$$

where d is the diameter of the contact circle at full load (assumed to be equal to the diameter of the residual impression in the surface). In nanoindentation testing, is it usual to find that the size of the residual impression is too small to be measured accurately with conventional techniques and instead, the depth of circle of contact is measured (the so-called "plastic depth" h_p as shown in Fig. 1.1) and the area of contact calculated using the known geometry of the indenter. For a spherical indenter, the area of contact is given by:

$$\begin{aligned} A &= \pi\left(2R_i h_p - h_p^2\right) \\ &\approx 2\pi R_i h_p \end{aligned} \tag{2.3.1b}$$

where the approximation is appropriate when the indentation depth is small compared to the radius of the indenter.

The mean contact pressure determined from Eq. 2.3.1a is based on measurements of the projected area of contact and is often called the "Meyer" hardness H. By contrast, the Brinell hardness number (BHN) uses the actual area of the curved surface of the impression and is found from:

$$\mathrm{BHN} = \frac{2P}{\pi D\left(D - \sqrt{D^2 - d^2}\right)} \tag{2.3.1c}$$

where D is the diameter of the indenter. The Brinell hardness is usually performed at a value for a/R (the indentation strain) of 0.4, a value found to be consistent with a fully developed plastic zone. The angle of a Vickers indenter (see Section 2.3.2 below) was chosen originally so as to result in this same level of indentation strain.

The use of the area of the actual curved surface of the residual impression in the Brinell test was originally thought to compensate for strain-hardening of the specimen material during the test itself. However, it is now more generally recognized that the Meyer hardness is a more physically meaningful concept.

Meyer found that there was an empirical size relationship between the diameter of the residual impression and the applied load, and this is known as Meyer's law:

$$P = kd^n \tag{2.3.1d}$$

In Eq. 2.3.1c, k and n are constants for the specimen material. It was found that the value of n was insensitive to the radius of the indenter and is related to the strain-hardening exponent x of the specimen material according to

$$n = x + 2 \tag{2.3.1e}$$

Values of n were found to be between 2 and 2.5, the higher the value applying to annealed materials, while the lower value applying to work-hardened materials (low value of x in Eq. 2.3.1d). It is important to note that Meyer detected a lower limit to the validity of Eq. 2.3.1c. Meyer fixed the lower limit of validity to an indentation strain of $a/R = 0.1$. Below this, the value of n was observed to increase — a result of particular relevance to nanoindentation testing.

2.3.2 Vickers Indenter

For a Vickers diamond pyramid indenter (a square pyramid with opposite faces at an angle of 136° and edges at 148° and face angle 68°), the Vickers diamond hardness, VDH, is calculated using the indenter load and the actual surface area of the impression. The VDH is lower than the mean contact pressure by ≈ 7%. The Vickers diamond hardness is found from:

$$VDH = \frac{2P}{d^2}\sin\frac{136°}{2} = 1.8544\frac{P}{d^2} \tag{2.3.2a}$$

with d equal to the length of the diagonal measured from corner to corner on the residual impression in the specimen surface. Traditionally, Vickers hardness is calculated using Eq. 2.3.2a with d in mm and P in kgf[§]. The resulting value is called the Vickers hardness and given the symbol HV. The mean contact pressure, or Meyer hardness, is found using the projected area of contact, in which case we have:

$$p_m = H = 2\frac{P}{d^2} \tag{2.3.2b}$$

There is a direct correspondence between HV and the Meyer hardness H.

$$HV = 94.5H \tag{2.3.2c}$$

[§] 1 kgf = 9.806 N

where H is expressed in MPa. In nanoindentation testing, the area of contact is found from a determination of the plastic depth h_p. The projected area of contact is given by:

$$\begin{aligned} A &= 4h_p{}^2 \tan^2 68 \\ &= 24.504h_p{}^2 \end{aligned} \qquad (2.3.2d)$$

2.3.3 Berkovich Indenter

The Berkovich indenter is used routinely for nanoindentation testing because it is more readily fashioned to a sharper point than the four-sided Vickers geometry, thus ensuring a more precise control over the indentation process. The mean contact pressure is usually determined from a measure of the "plastic" depth of penetration, h_p in (see Fig. 1.3), such that the projected area of the contact is given by:

$$A = 3\sqrt{3}h_p{}^2 \tan^2 \theta \qquad (2.3.3a)$$

which for $\theta = 65.27°$, evaluates to:

$$\begin{aligned} A &= 24.494h_p{}^2 \\ &\approx 24.5h_p{}^2 \end{aligned} \qquad (2.3.3b)$$

and hence the mean contact pressure, or hardness, is:

$$H = \frac{P}{24.5h_p{}^2} \qquad (2.3.3c)$$

The original Berkovich indenter was designed to have the same ratio of actual surface area to indentation depth as a Vickers indenter and had a face angle of 65.0333°. Since it is customary to use the mean contact pressure as a definition of hardness in nanoindentation, Berkovich indenters used in nanoindentation work are designed to have the same ratio of projected area to indentation depth as the Vickers indenter in which case the face angle is 65.27°.

For both the Vickers and the Berkovich indenters, the representative strain within the specimen material is approximately 8% (see Section 1.3).

2.3.4 Cube Corner Indenter

The Berkovich and Vickers indenters have a relatively large face angles, which ensures that deformation is more likely to be described by the expanding cavity model rather than slip-line theory, which is equivalent to saying that the stresses beneath the indenter are very strongly compressive. In some instances, it is desirable to indent a specimen with more of a cutting action, especially when intentional radial and median cracks are required to measure fracture toughness. A cube corner indenter offers a relatively acute face angle that can be beneficial in these circumstances. Despite the acuteness of the indenter, it is still possible to perform nanoindentation testing in the normal manner and the expression for the projected area of contact is the same as that for a Berkovich indenter where in this case the face angle is $\theta = 35.26°$:

$$\begin{aligned} A &= 3\sqrt{3}h_p^{\ 2}\tan^2\theta \\ &= 2.60h_p^{\ 2} \end{aligned} \tag{2.3.4a}$$

2.3.5 Knoop Indenter

The Knoop indenter is similar to the Vickers indenter except that the diamond pyramid has unequal length edges, resulting in an impression that has one diagonal with a length approximately seven times the shorter diagonal.[3] The angles for the opposite faces of a Knoop indenter are 172.5° and 130°. The Knoop indenter is particularly useful for the study of very hard materials because the length of the long diagonal of the residual impression is more easily measured compared to the dimensions of the impression made by Vickers or spherical indenters.

As shown in Fig. 2.4, the length d of the longer diagonal is used to determine the projected area of the impression. The Knoop hardness number is based upon the projected area of contact and is calculated from:

$$KHN = \frac{2P}{a^2\left[\cot\frac{172.5}{2}\tan\frac{130}{2}\right]} \tag{2.3.5a}$$

For indentations in highly elastic materials, there is observed a substantial difference in the length of the short axis diagonal for a condition of full load compared to full unload. Marshall, Noma, and Evans[4] likened the elastic recovery along the short axis direction to that of a cone with major and minor axes and applied elasticity theory to arrive at an expression for the recovered indentation size in terms of the geometry of the indenter and the ratio H/E:

$$\frac{b'}{a'} = \frac{b}{a} - \alpha\frac{H}{E} \tag{2.3.5b}$$

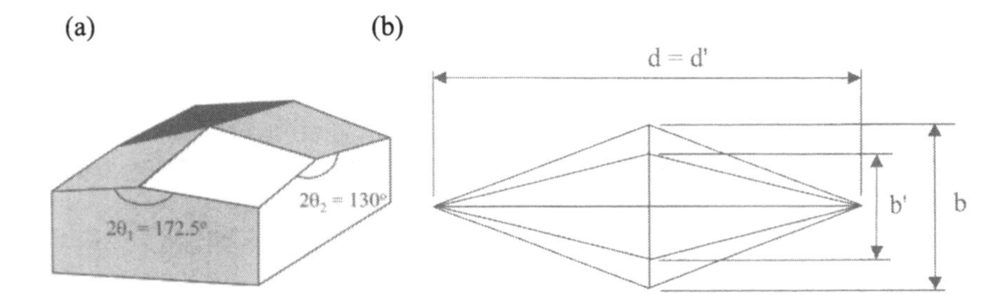

Fig. 2.4 (a) Geometry of a Knoop indenter. (b) The length of the long diagonal of the residual impression remains approximately the same from full load to full unload. The size of the short diagonal reduces from b to b' due to elastic recovery during unloading.

In Eq. 2.3.5b, α is a geometry factor found from experiments on a wide range of materials to be equal to 0.45. The ratio of the dimension of the short diagonal b to the long diagonal a at full load is given by the indenter geometry and for a Knoop indenter, $b/a = 1/7.11$. The primed values of a and b are the lengths of the long and short diagonals after removal of load. Since there is observed to be negligible recovery along the long diagonal, then $a' \approx a$. When H is small and E is large (e.g. metals), then $b' \approx b$ indicating negligible elastic recovery along the short diagonal. When H is large and E is small (e.g. glasses and ceramics), then $b' \ll b$. Using measurements of the axes of the recovered indentations, it is possible to estimate the ratio E/H for a specimen material using Eq. 2.3.5b.

2.4 Load-Displacement Curves

The principal goal of nanoindentation testing is to extract elastic modulus and hardness of the specimen material from experimental readings of indenter load and depth of penetration. In a typical test, load and depth of penetration are recorded as load is applied from zero to some maximum and then from maximum load back to zero. If plastic deformation occurs, then there is a residual impression left in the surface of the specimen. Unlike conventional indentation hardness tests, the size (and hence the projected contact area) of the residual impression for nanoindentation testing is too small to measure accurately with optical techniques. The depth of penetration together with the known geometry of the indenter provides an indirect measure of the area of contact at full load, from which the mean contact pressure, and thus hardness, may be estimated. When load is removed from the indenter, the material attempts to regain its original shape, but it prevented from doing so because of plastic deformation. However, there is some degree of recovery due to the relaxation of elastic strains within the material. An analysis of the initial portion of this elastic unloading response gives an estimate of the elastic modulus of the indented material.

The form of the compliance curves for the most common types of indenter are very similar and is shown in Fig. 2.5. For a spherical indenter, it will be shown in Chapter 5 that the relationship between load and penetration depth for the loading portion for an elastic–plastic contact is given by:

$$h = \frac{1}{2}\left(\frac{P}{\pi R_i H} + \frac{3}{4}\frac{\sqrt{P\pi H}}{\beta E^*}\right) \tag{2.4a}$$

For the elastic unloading, we have from Eq. 1.2g:

$$h = \left[\frac{3}{4E^* R'^{1/2}}\right]^{2/3} P^{\frac{2}{3}} \tag{2.4b}$$

For a Berkovich indenter, it will be shown in Chapter 5 that the expected relationship between load and depth for an elastic–plastic contact is given by:

$$h = \sqrt{P}\left[\left(3\sqrt{3}H\tan^2\theta\right)^{-\frac{1}{2}} + \left[\frac{2(\pi-2)}{\pi}\right]\frac{\sqrt{H\pi}}{2\beta E^*}\right] \tag{2.4c}$$

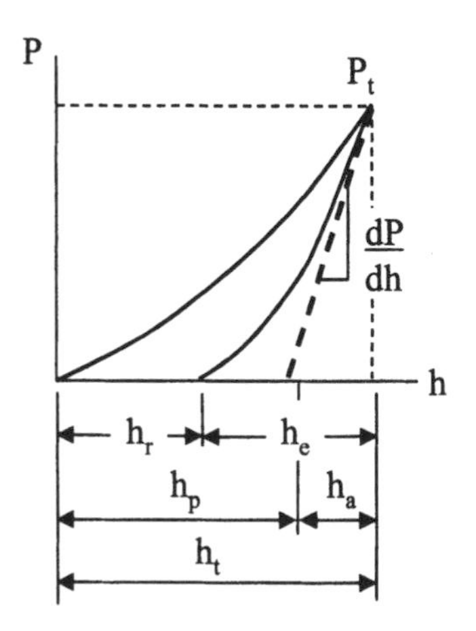

Fig. 2.5 Compliance curves, loading and unloading, from a nanoindentation experiment with maximum load P_t and depth beneath the specimen free surface h_t. The depth of the contact circle h_p and slope of the elastic unloading dP/dh allow specimen modulus and hardness to be calculated. h_r is the depth of the residual impression, and h_e is the displacement associated with the elastic recovery during unloading.

Upon elastic unloading we have from Eq. 1.2m:

$$h = \sqrt{P}\left(\frac{\pi}{2E^*}\right)^{\frac{1}{2}}\left(\frac{\pi}{3\sqrt{3}}\right)^{\frac{1}{4}}\frac{1}{\tan\theta'} \tag{2.4d}$$

where in Eqs. 2.4b and 2.4d the quantities R' and θ' are the combined radii and angle of the indenter and the shape of the residual impression in the specimen surface. The dependence of depth on the square root of the applied load in Eqs. 2.4a to 2.4d is of particular relevance. This relationship is often used in various methods of analysis to be described in Chapter 3.

In subsequent chapters, the methods by which elastic modulus and hardness values are obtained from experimental values of load and depth are described along with methods of applying necessary corrections to the data. In most cases, methods of analysis rely on the assumption of an elastic–plastic loading followed by an elastic unloading — with no plastic deformation (or "reverse" plasticity) occurring during the unloading sequence.

The indentation modulus is usually determined from the slope of the unloading curve at maximum load. Eq. 2.4e shows that the indentation modulus (here expressed as E^*) as a function of dP/dh and the area of contact.

$$E^* = \frac{1}{2}\frac{\sqrt{\pi}}{\sqrt{A}}\frac{dP}{dh} \tag{2.4e}$$

The indentation hardness is calculated from the indentation load divided by the contact area. The contact area in turn is determined from the value of h_p (see Fig. 1.3) and the known geometry of the indenter (Table 1.1). The value for h_p is found by an analysis of the load-displacement data (see Fig. 2.5).

Variations on the basic load - unload cycle include partial unloading during each loading increment, superimposing an oscillatory motion on the loading, and holding the load steady at a maximum load and recording changes in depth. These types of tests allow the measurement of viscoelastic properties of the specimen material.

In practice, nanoindentation testing is performed on a wide variety of substances, from soft polymers to diamond-like carbon thin films. The shape of the load-displacement curve is often found to be a rich source of information, not only for providing a means to calculate modulus and hardness of the specimen material, but also for the identification of non-linear events such as phase transformations, cracking, and delamination of films. Fig. 2.6 shows a schematic of some of the more commonly observed phenomena. It should be noted that in many cases the permanent deformation or residual impression is not the result of plastic flow but may involve cracking or phase changes within the specimen.

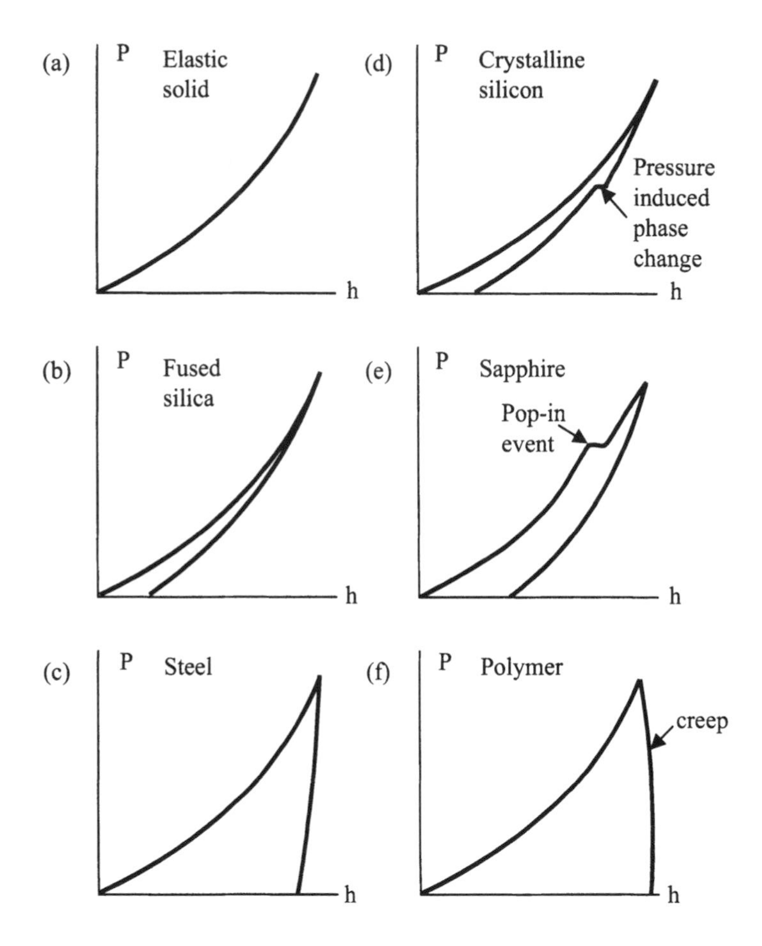

Fig. 2.6 Schematic examples of load-displacement curves for different material responses and properties. (a) Elastic solid, (b) brittle solid, (c) ductile solid, (d) crystalline solid, (e) brittle solid with cracking during loading, and (f) polymer exhibiting creep.

2.5 Experimental Techniques

Despite the mature evolution of nanoindentation test instruments, the process of undertaking such a test requires considerable experimental skill and resources. Such tests are extremely sensitive to thermal expansion from temperature changes and mechanical vibration during testing. It is necessary to ensure that the specimen and the instrument are in thermal equilibrium. For example, handling the specimen or the indenter requires a reasonable delay before beginning the indentation so that errors are not introduced into the displacement measurements by virtue of thermal expansion or contraction during the test. Should there be any long-term thermal drifts, then these should be quantified and the appropriate correction made (see Chapter 4). In this section, specified matters requiring attention in practical indentation tests are summarized and commented upon.

2.5.1 Instrument Construction and Installation

The nanoindentation instrument should be insulated against temperature variation, vibration, and acoustic noise in normal laboratory conditions. A specially designed enclosure designed to reduce thermal and electrical interference to a minimum is usually supplied by the manufacturer as part of the installation. For very low penetration depths (<100 nm) an active vibration mounting may be required.

The loading column and base of a nanoindentation instrument should be of heavy construction so as to act as a seismic mass (to reduce the transmission of mechanical vibration) and to have a very high compliance (to minimize the effect of reaction forces on the displacement readings). The indenter is typically mounted on a shaft that is made from stiff, yet lightweight material to minimize compliance and maximize the resonant frequency of the system (the latter condition being important for dynamic indentation testing).

Specimens are typically mounted on a metal base with wax or mounting adhesive. The specimen holder is in turn placed on a stage. Stage movement is usually controlled by motorized axes that have a resolution, or step size, of less than 0.5 μm. Such fine positioning is usually needed to allow indentations to be made on very small features such as grains in a ceramic or conductive pads in an integrated circuit. Stage movement is usually servo controlled with proportional, integral, and derivative gains that can be set to allow for the most precise positioning. Optical rotary or linear track encoders are usually employed. The encoder is usually mounted on the lead screw of the axis drive and a high-quality ball nut drives converts rotary motion to linear motion of the stage. Backlash in this nut is negligible. The operating software allows automatic positioning of the specimen beneath the indenter.

Load is typically applied to the indenter shaft by an electromagnetic coil or the expansion of a piezoelectric element. Displacements are usually measured using either a changing capacitance or inductance signal. Most nanoindentation instruments are load controlled, that is, a commanded force is applied and the resulting displacement is read. Particular features of some commercially available indentation instruments are described in Chapter 11.

2.5.2 Indenters

Diamond indenters are very hard, but also very brittle and can easily be chipped or broken (see Fig. 2.7). The mechanical properties of diamond differ according to the orientation of the measurement due to the crystalline nature of the diamond structure. Literature values for modulus range from about 800 GPa to 1200 GPa. A value of ≈1000 GPa is usually used in the analysis of nanoindentation test data with a Poisson's ratio of 0.07.

Fig. 2.7 Brittle failure of a 2 μm radius sphero-conical indenter. (Courtesy CSIRO).

The indenter must be absolutely clean and free from any contaminants. Diamond indenters are most effectively cleaned by pressing them into a block of dense polystyrene. The chemicals in the polystyrene act as a solvent for any contaminants and the polystyrene itself offers a mechanical cleaning action that is not likely to fracture the indenter. The indenter itself should be attached to the indenter shaft firmly and in such a manner so as to minimize its compliance — this is often a matter for the manufacturer of the instrument.

The choice of indenter is important and depends upon the information one wishes to obtain from the indentation test. The representative strain in the specimen material, for geometrically similar indentations such as that made by Vickers and Berkovich indenters, depends solely on the effective cone angle of the indenter. The sharper the angle, the greater the strain. According to Tabor, the representative strain for a conical indenter is given by:

$$\varepsilon = 0.2 \cot \alpha \tag{2.5.2a}$$

which for a Berkovich and Vickers indenter, evaluates to about 8%. If larger strains are required, say, for example, to induce cracking or other phenomena, then a sharper tip may be required. The representative strain for a cube corner indenter evaluates to about 22%. Indentations made with sharp indenters induce plasticity from the moment of contact (neglecting any tip-rounding effects). This may be desirable when testing very thin films in which the hardness of the film, independent of the substrate is required.

Spherical indenters offer a gradual transition from elastic to elastic–plastic response. The representative strain varies as the load is applied according to:

$$\varepsilon = 0.2\frac{a}{R} \tag{2.5.2b}$$

It is important that when measuring hardness using a spherical indenter, a fully developed plastic zone is obtained. In metals, this usually corresponds to a value for a/R of greater than 0.4. The changing strain throughout an indentation test with a spherical indenter enables the elastic and elastic–plastic properties of the specimen to be examined along with any strain-hardening characteristics.

2.5.3 Specimen Mounting

Test specimens must be presented square on to the axis of the indenter and held firmly with the absolute minimum of compliance in the mounting. Typically, a specimen is mounted onto a hardened base or specimen mount using a very thin layer of glue. Holding devices such as magnets, vacuum chucks or spring clamps may be used.

The working range of a typical nanoindentation instrument is usually in the order of microns and so the specimen surface, if many indentations are to be made, is required to be parallel with the axis of translation. A departure from parallelism of about 25 μm over a 10 mm traverse can typically be tolerated by an instrument.

If a polishing technique is employed to prepare a surface, the properties of the surface material may be altered (see Section 4.11). Indentations placed on scratches, inclusions, or voids will give unpredictable results.

Fig. 2.8 Example of a fused silica specimens of diameter ≈ 10 mm mounted on a cylindrical hardened steel specimen mount and placed into position on a servo-motor-driven X-Y positioning stage. In this instrument, magnets firmly clamp the specimen mount to the stage (Courtesy CSIRO).

2.5.4 Working Distance and Initial Penetration

A typical nanoindentation instrument has a limited range of displacement over which the indentation depth may be measured. It is therefore necessary to ensure that the full range of the depth measurement system is available for measuring penetration depth into the specimen and not used for bringing the indenter into contact with the surface from its initial parked position. Usually, the measurement head of the instrument is allowed to translate vertically in coarse steps until the indenter is within 100 μm or so of the specimen surface. This is called the "working distance" and ensures that most of the available high-resolution displacement occurs during the final approach to the surface and the subsequent penetration into the specimen.

Once the measurement head has been set so that the indenter is at the working distance, it is then necessary to bring the indenter down to touch the surface with the minimum possible contact force. This becomes the reference position for any subsequent displacement readings. The minimum contact force is a very important measurement of performance for a nanoindentation instrument. No matter how small the minimum contact force is, it will result in some penetration into the specimen surface that has to be corrected for in the final analysis. Of course, if the initial penetration is too large, then it is possible that the indenter will penetrate past the surface layer or film desired to be measured. In many respects, the specification of the minimum contact force is the parameter that distinguishes a nanoindentation instrument from a microindentation instrument.

Some nanoindentation instruments apply the initial contact force by bringing the indenter down at a very small velocity until a preset initial contact force is measured by a separate force sensor. Others monitor the "stiffness" of the contact by oscillating the indenter shaft and noting when the oscillations undergo a sudden reduction in amplitude. The process can be automated so that it becomes independent of the operator. The initial contact force is typically in the range of 5 μN or less.

2.5.5 Test Cycles

A typical nanoindentation test cycle consists of an application of load followed by an unloading sequence — but there are many variations. Load may be applied continuously until the maximum load is reached, or as a series of small increments. At each load increment, a partial unloading may be programmed that provides a measurement of stiffness of the contact (dP/dh), which is important for measuring changes in modulus or hardness with penetration depth. Contact stiffness may also be found by superimposing a small oscillatory motion onto the load signal.

The indentation instrument may be set into either load or depth control. In load control, the user specifies the maximum test force (usually in mN) and the number of load increments or steps to use. The progression of load increments may be typically set to be a square root or linear progression. A square root progression attempts to provide equally spaced displacement readings. In depth control, the user specifies a maximum depth of penetration. However, it should be noted that most nanoindentation instruments are inherently load-controlled devices and when operating under a depth control mode, they typically apply small increments of force until the required depth has been reached.

It is customary for a nanoindentation instrument to allow for a dwell or hold period at each load increment and at maximum load. The dwell settings at each load increment allow the instrument and specimen to stabilize before depth and load readings are taken.

Hold period data at maximum load can be used to measure creep within the specimen or thermal drift of the apparatus during a test. Hold measurements for the purposes of thermal drift are probably best carried out at the end of the indentation test, at a low load, so as to minimize any effects from creep within the specimen. Figure 2.9 summarizes the loading, hold and unloading periods in a typical test cycle.

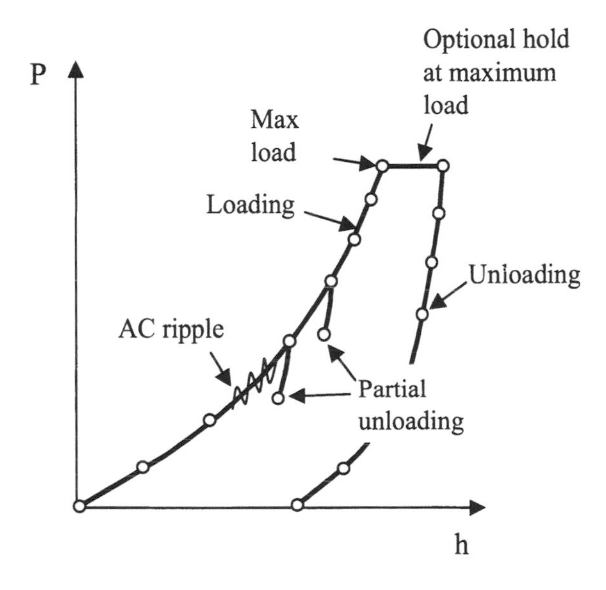

Fig. 2.9 Various components of a nanoindentation test cycle.

References

1. E.S. Berkovich, "Three-faceted diamond pyramid for micro-hardness testing," Ind. Diamond Rev. 11 127, 1951, pp. 129–133.
2. B.R. Lawn, *Fracture of Brittle Solids*, 2nd Ed., Cambridge University Press, Cambridge, 1993.
3. F. Knoop, C.G. Peters, and W.B. Emerson, "A sensitive pyramidal-diamond tool for indentation measurements," Research Paper 1220, Journal of Research, National Bureau of Standards, 23 1, 1939.
4. D.B. Marshall, T. Noma, and A.G. Evans, "A simple method for determining elastic-modulus-to-hardness ratios using Knoop indentation measurements," J. Am. Ceram. Soc. 65, 1980, pp. C175–C176.

Chapter 3
Analysis of Nanoindentation Test Data

3.1 Analysis of Indentation Test Data

As described in Chapter 2, estimations of both elastic modulus and hardness of the specimen material in a nanoindentation test are obtained from load versus penetration measurements. Rather than a direct measurement of the size of residual impressions, contact areas are instead calculated from depth measurements together with a knowledge of the actual shape of the indenter. For this reason, nanoindentation testing is sometimes referred to as depth-sensing indentation testing. In this chapter, methods of the analysis of load-displacement data that are used to compute hardness and modulus of test specimens are presented in detail. It is an appropriate introduction to first consider the case of a cylindrical punch indenter — even though this type of indenter is rarely used for this type of testing, its response illustrates and introduces the theory for the more complicated cases of spherical and pyramidal indenters.

3.2 Analysis Methods

3.2.1 Cylindrical Punch Indenter

Consider the case of a cylindrical flat punch indenter that has an elastic–plastic load displacement response as shown in Fig. 3.1. Figure 3.1 (a) shows the displacements for the elastic–plastic contact at full load P_t and the displacements at full unload. The unloading response is assumed to be fully elastic. Elastic displacements can be calculated using Eq. 3.2.1a:

$$P = 2aE^*h \tag{3.2.1a}$$

Putting h equal to the displacement u_z at $r = 0$ and by taking the derivative dP/dh, we can arrive at an expression for the slope of the unloading curve:

$$\frac{dP}{dh} = 2E^*a \tag{3.2.1b}$$

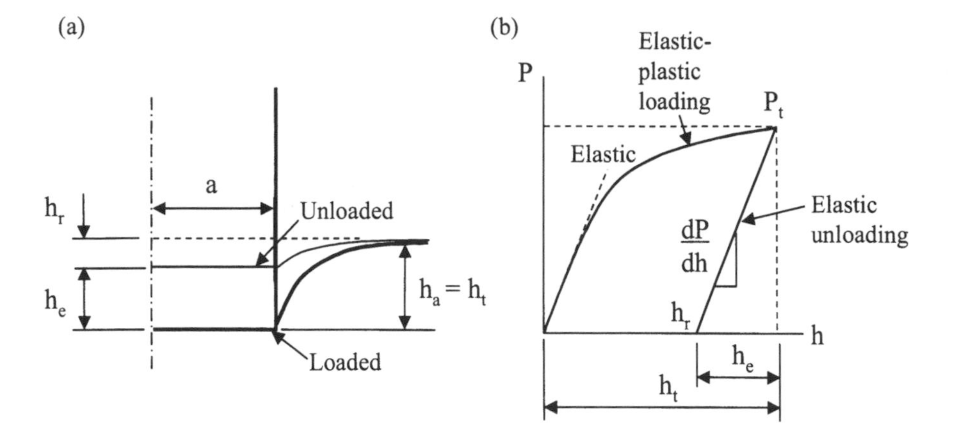

Fig. 3.1 (a) Schematic of indenter and specimen surface geometry at full load and full unload for cylindrical punch indenter. (b) Load versus displacement for elastic–plastic loading followed by elastic unloading. h_r is the depth of the residual impression, h_t is the depth from the original specimen surface at maximum load P_t, h_e is the elastic displacement during unloading, and h_a is the distance from the edge of the contact to the specimen surface, here equal to h_t for cylindrical indenter (after reference 1).

In Eqs. 3.2.1a and 3.2.1b, a is the contact radius, which, for the case of a cylindrical punch, is equal to the radius of the indenter. Expressing this in terms of the contact area:

$$\frac{dP}{dh} = 2E^* \frac{\sqrt{A}}{\sqrt{\pi}} \tag{3.2.1c}$$

Pharr, Oliver, and Brotzen[2] show that Eq. 3.2.1c applies to all axial-symmetric indenters. Equation 3.2.1c shows that the slope of the unloading curve is proportional to the elastic modulus and may be calculated from the known radius of the punch. As shown in Fig. 3.1, h_e is the displacement for the elastic unloading. Thus, the slope of the unloading curve is also given by:

$$\frac{dP}{dh} = \frac{P_t}{h_e} \tag{3.2.1d}$$

Now, for a cylindrical indenter, there is no need for an estimation of the size of the contact area from depth measurements because it is equal to the radius of the indenter. However, the situation becomes quite complicated when this is not the case, such as for the Berkovich indenter.

3.2.2 Conical Indenter — Cylindrical Punch Approximation

As can be seen from Fig. 3.1, the slope of the elastic unloading curve for a cylindrical punch indenter is linear. Doerner and Nix[3] observed that for tests with a Berkovich indenter, the *initial* unloading curve appeared to be linear for a wide range of test materials. They then applied cylindrical punch equations to the unloading data to determine the size of the contact from depth measurements. Their analysis considered the case of a conical indenter and assumed that the actual pyramidal geometry had only a small effect on the final result.[2]

Thus, consider the elastic–plastic loading and elastic unloading of a specimen with a conical indenter. The shape of the surface for the sequence is shown in Fig. 3.2 (a). As the indenter is unloaded from full load, the contact radius remains fairly constant (due to a fortuitous combination of the geometry of the deformation and the shape of the indenter) until the surface of the specimen no longer conforms to the shape of the indenter. Thus, for the initial part of the unloading, if the contact radius is assumed to be constant, the unloading curve is linear. Unloading from the fully loaded impression for a cone is therefore similar to that seen for the elastic unloading of a cylindrical punch.

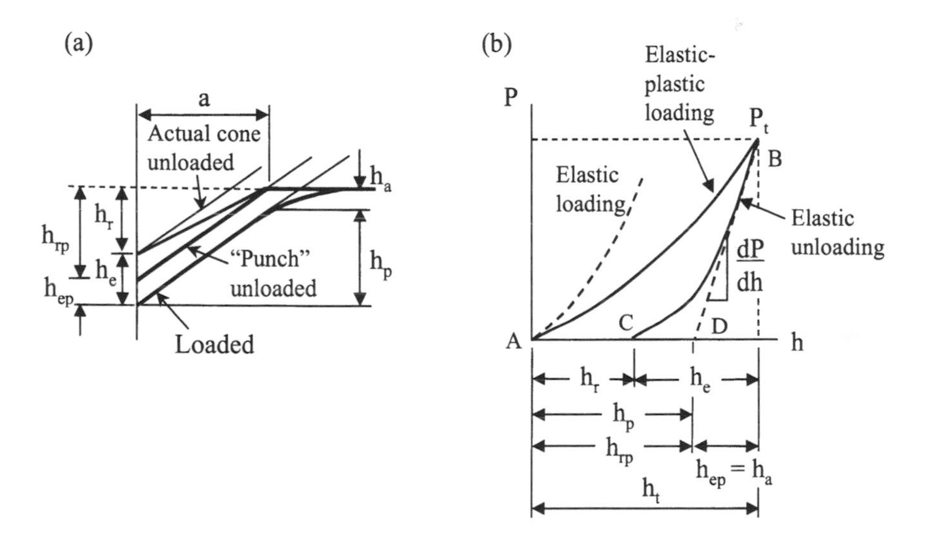

Fig. 3.2 (a) Schematic of indenter and specimen surface geometry at full load and full unload for conical indenter. (b) Load versus displacement for elastic–plastic loading followed by elastic unloading. h_r is the depth of the residual impression, h_{rp} is the depth of the residual impression for an equivalent punch, h_t is the depth from the original specimen surface at maximum load P_t, h_e is the elastic displacement during unloading of the actual cone, h_{ep} is the elastic displacement for an equivalent punch, and h_a is the distance from the edge of the contact to the specimen surface at full load (after reference 1).

Doerner and Nix[3] thus use Eq. 3.2.1c, with A being the contact area of the punch, to obtain the depth of the edge of the circle of contact h_a, and hence h_p. The radius of the circle of contact at full load is obtained from the slope of the initial unloading.

With reference to Fig. 3.2 (b), the cone acts like a punch in terms of its "constant" area of contact, since the initial unloading is appears to be linear. If the area of contact remained constant during the entire unloading, then the unloading curve would be linear: from P_t to $P = 0$. The unloading curve associated with a punch that has the slope dP/dh at P_t would be that which is extrapolated to zero load, that is, the path BD in Fig. 3.2 (b). If unloading were to take place along this line, then the elastic displacement of this imaginary punch would be the distance h_{ep} (the subscript "p" denoting "punch"). The cone actually travels the path BC during unloading, which is a distance h_e, leaving a residual impression of depth h_r. Now, with reference to Fig. 3.2 (a), it is easy to see that for a cone that acts like a punch (by having a constant radius of circle of contact during unloading) and unloads through a distance h_{ep}, the distance h_{ep} is equal to the distance h_a *that exists at full load,* which is where an estimate of h_a for the actual cone is required. As a consequence, the distance h_{rp}, which is the intercept of the unloading curve for the punch, is equal to h_p. Thus:

$$\begin{aligned} h_t - h_{rp} &= h_s \\ h_{rp} &= h_t - h_s \\ &= h_p \end{aligned} \qquad (3.2.2a)$$

Equation 3.2.2a indicates that the depth h_{rp}, and hence h_p, can be obtained from the intercept of the linear unloading curve with the displacement axis. Once h_p is known, then the area of the contact can be calculated and the hardness and elastic modulus determined from the geometry of the indenter. For example, for a Vickers or Berkovich indenter, the relationship between the projected area A_p of the indentation and the distance h_p is:

$$A_p = 24.5 h_p^{\,2} \qquad (3.2.2b)$$

Thus:

$$\frac{dP}{dh} = 2 h_p E^* \sqrt{\frac{24.5}{\pi}} \qquad (3.2.2c)$$

The displacement h_p is found from the intercept of the linear unloading curve with the displacement axis. The actual cone and the imaginary punch meet at P_t. It is assumed that the initial unloading can be extrapolated to zero load which provides a measure of h_{ep}. Due to the geometrical similarity of this punch-like cone, $h_{ep} = h_a$, and thus h_p for the actual cone is determined. Values of H and E can be calculated from the maximum load P_t divided by the projected area (Eq. 3.2.2b) A_p and from Eq 3.2.2c, respectively.

3.2.3 Spherical Indenter

Historically, the cylindrical punch equations were applied to data obtained with a Berkovich indenter. Oliver and Pharr[4] later noted that the unloading response for many materials tested with a Berkovich indenter was curved rather than linear. This lead these workers to use a power law relationship of a cone rather than the linear relationship of a cylindrical punch to the unloading data. Oliver and Pharr considered the case of a Berkovich indenter but noted that the method could be readily extended to other indenters. Field and Swain[5] applied the elastic equations of contact directly to the unloading data (rather than the slope of a fit to the data) for the case of a spherical indenter. This method can also be used for different types of indenter.

Consider the loading of an initially flat specimen with a spherical indenter. Upon loading, there may be an initial elastic response at low loads followed by elastic and plastic deformations within the specimen material at higher loads.

With reference to Fig. 3.3, the depth of penetration of a rigid spherical indenter beneath the original specimen free surface is h_t at full load P_t. When the load is removed, assuming no reverse plasticity, the unloading is elastic and at complete unload, there is a residual impression of depth h_r.

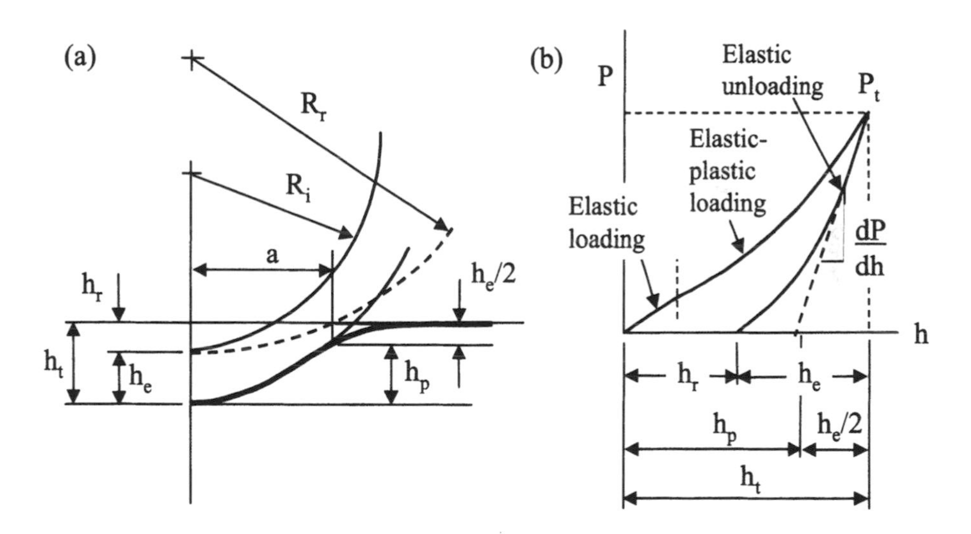

Fig. 3.3 (a) Geometry of loading a preformed impression of radius R_r with a rigid indenter radius R_i. (b) Compliance curve (load vs displacement) for an elastic–plastic specimen loaded with a spherical indenter showing both loading and unloading response. Upon loading, there is an initial elastic response followed by elastic–plastic deformation. Upon complete unload, there is a residual impression of depth h_r (after reference 1).

If the load P_t is reapplied, then the reloading is elastic through a distance $h_e = h_t - h_r$ according to the "Hertz" equation:

$$P = \frac{4}{3} E^* R^{1/2} h_e^{3/2} \tag{3.2.3a}$$

It should be borne in mind that the quantity h_e in Eq. 3.2.3a is, strictly speaking, the load-point displacement, which is only equal to the displacement of the specimen surface along the axis of symmetry (r = 0) for the special case of a perfectly rigid indenter. However, E^* is assigned to the specimen, then it is possible to consider the indentation to be occurring with a perfectly rigid indenter and to extract the actual specimen modulus E at the end of the analysis using the known values of mechanical properties E' and v' for the indenter. Furthermore, it should be noted that since the elastic unloading/reloading involves the deformation of the preformed residual impression, R in Eq. 3.2.3a is the relative radius of curvature of the residual impression R_r and the indenter R_i and is given by:

$$\frac{1}{R} = \frac{1}{R_i} - \frac{1}{R_r} \tag{3.2.3b}$$

Note that the center of curvature for the residual impression is on the same side of the surface as the indenter, hence the presence of the negative sign in Eq. 3.2.3b. The radius of the residual impression serves to increase the effective radius of the indenter. It will be shown that it is not necessary to measure the radius of the residual impression R_r to perform the analysis.

There are two important matters to consider at this point. First, the chordal diameter of the residual impression may be assumed to be identical to that of the circle of contact at full load (owing to a fortuitous combination of the geometry of the deformation and the shape of the indenter). That is, during an imagined elastic reloading of the residual impression, the radius of the circle of contact between the indenter and the specimen moves outward (and downward) until it meets the edge of the residual impression, by which time the load has reached P_t. Second, since the loading/unloading from h_r to h_t is elastic, the Hertz equations show that depth of the circle of contact beneath the specimen free surface is half of the elastic displacement h_e. That is, the distance from the specimen free surface (at full unload) to the depth of the radius of the circle of contact at full load is $h_a = h_e/2$. With reference to Fig. 3.3:

$$\begin{aligned} h_t &= h_p + h_a \\ h_p &= h_t - \frac{h_e}{2} \end{aligned} \tag{3.2.3c}$$

The multiple-point and the single-point unload methods are concerned with the determination of the quantity h_p, often referred to as the "plastic depth," and is the distance from the circle of contact to the maximum penetration depth. Once h_p is known, the resulting radius of circle of contact is then determined by simple geometry from which the area of the contact is calculated and used to determine the mean contact pressure or hardness value H:

$$H = \frac{P}{A} \tag{3.2.3d}$$

where A is the area of contact given by πa^2 with the term "a" being the radius of the circle of contact at $P = P_t$. Elastic modulus is determined from the slope of the unloading curve or by the Hertz equations directly.

3.2.3.1 Multiple-Point Unload Method

The multiple-point unload method** uses the slope of the initial portion of the unloading curve to determine the depth of the circle of contact h_a and hence h_p. The slope of the elastic unloading, for the case of a spherical indenter, is given by the derivative of Eq. 3.2.3a with respect to h:

$$\frac{dP}{dh} = 2E^* R^{1/2} h_e^{\ 1/2} \tag{3.2.3.1a}$$

The quantity dP/dh is sometimes referred to as the contact stiffness and given the symbol S. Substituting Eq. 3.2.3.1a into Eq. 3.2.3a, we have:

$$P = \frac{2}{3}\frac{dP}{dh} h_e \tag{3.2.3.1b}$$

Thus,

$$h_e = \frac{3}{2} P \frac{dh}{dP} \tag{3.2.3.1c}$$

With reference to Fig. 3.3, the unloading from h_t to h_r is assumed to be elastic, and for a spherical indenter, the Hertz equations show that the depth of the circle of contact h_a beneath the specimen free surface is half of the elastic displacement h_e, that is:

$$h_a = \frac{h_e}{2} \tag{3.2.3.1d}$$

Thus:

$$h_a = \left[\frac{3}{4}\right] \frac{P_t}{dP/dh} \tag{3.2.3.1e}$$

Once a value for h_a is obtained, the plastic depth h_p can be found from Eq. 3.2.3c. The radius of the circle of contact can then be found from geometry:

** This is the "Oliver and Pharr" method applied to a spherical indenter.

$$a = \sqrt{2R_i h_p - h_p{}^2} \approx \sqrt{2R_i h_p} \tag{3.2.3.1f}$$

and the hardness computed from the load divided by the area of contact. The approximation in Eq. 3.2.3.1f is precisely the same as that underlying the Hertz equations and is equivalent to the restriction that the indentations are to be small — i.e. h_p <<a. For a rigid spherical indenter, Hertz showed that the elastic displacement is given by:

$$h_e = \frac{a^2}{R} \tag{3.2.3.1g}$$

whereupon Equation 3.2.3.1a becomes:

$$\frac{dP}{dh} = 2E^* R^{1/2} \frac{a}{R^{1/2}} = 2E^* a \tag{3.2.3.1h}$$

Note that the relative radius of the indenter and specimen R has been cancelled out in Eq. 3.2.3.1h. The combined modulus of the system can thus be determined from the slope of the initial unloading:

$$E^* = \frac{dP}{dh}\frac{1}{2a} = \frac{1}{2}\frac{dP}{dh}\frac{\sqrt{\pi}}{\sqrt{A}} \tag{3.2.3.1i}$$

where $A = \pi a^2$, the area of contact. Equation 3.2.3.1i should be compared with Eq. 3.2.1c and is a general relationship that applies to all axial-symmetric indenters with a smooth profile.[2] The multiple-point unload method is most well known when it is applied to indentations performed with a three-sided Berkovich indenter, although, as shown above, it is equally applicable to the case of spherical indenters.

3.2.3.2 Single-Point Unload Method

In contrast to the multiple-point unload method, the single-point unload method†† uses a single unload point together with the Hertz equations directly (rather than the derivative) as the basis for estimating the plastic depth and unloading characteristics, and hence hardness and modulus.

In terms of the radius of the circle of contact, the Hertz equation, Eq. 3.2.3a, can be expressed:

†† The "Field and Swain" method.

$$P = \frac{4}{3}E^{*}ah_{e} \tag{3.2.3.2a}$$

Since the unloading from h_t to h_r is elastic, the depth of the circle of contact beneath the specimen free surface is half of the elastic displacement h_e:

$$\begin{aligned} h_p &= h_t - \frac{h_e}{2} \\ &= h_t - \frac{h_t - h_r}{2} \\ &= \frac{h_t + h_r}{2} \end{aligned} \tag{3.2.3.2b}$$

The depth h_t is given directly by the instrument. The depth of the residual impression h_r can be found by a measurement of load and displacement at a reduced or a partial unload P_s from a higher load P_t and forming the ratio of the elastic displacements thus:

$$\begin{aligned} h_e = h_t - h_r &= \left[\left(\frac{3}{4E^{*}}\right)^{\frac{2}{3}}\frac{1}{R^{\frac{1}{3}}}\right]P_t^{2/3} \\ h_s - h_r &= \left[\left(\frac{3}{4E^{*}}\right)^{\frac{2}{3}}\frac{1}{R^{\frac{1}{3}}}\right]P_s^{2/3} \\ \frac{h_t - h_r}{h_s - h_r} &= \left(\frac{P_t}{P_s}\right)^{\frac{2}{3}} \\ h_r &= \frac{h_s\left(P_t/P_s\right)^{2/3} - h_t}{\left(P_t/P_s\right)^{2/3} - 1} \end{aligned} \tag{3.2.3.2c}$$

The plastic depth h_p and also the elastic displacement h_e can now be calculated from Eq. 3.2.3.2b. From geometry, the radius of the circle of contact at full load (at depth h_p) is given by Eq. 3.2.3.1f. The hardness H is thus computed from Eq. 3.2.3d. The modulus can be calculated using Eq. 3.2.3.2a using a calculated from Eq. 3.2.3.1f at $P = P_t$.

3.2.4 Berkovich Indenter

The Berkovich indenter has a face angle of $\theta = 65.27°$, which gives the same projected area-to-depth ratio as the Vickers indenter. It should be noted that the original Berkovich indenter[6] had a slightly different angle, that of 65.03°. This latter angle gives the same *actual* area to depth ratio as a Vickers indenter. However, it is generally agreed that the Vickers hardness scale (which uses the actual rather than the projected area of contact) is less physically meaningful than hardness values computed using the projected area of contact — which translates directly into mean contact pressure beneath the indenter.

For a Berkovich indenter, the relationship between the projected area A of the indentation and the depth h_p beneath the contact is:

$$\begin{aligned} A &= 3\sqrt{3}h_p^{\,2}\tan^2 65.27 \\ &= 24.5h_p^{\,2} \end{aligned} \tag{3.2.4a}$$

Once h_p is found, then the projected area of contact is thus calculated and the hardness computed from Eq. 3.2.3d. Elastic modulus can be found from an analysis of the slope of the initial unloading in a manner similar to that described above for the spherical indenter.

It is convenient to examine the details of the method with reference to an axial-symmetric cone rather than the actual non-symmetric pyramidal indenter. It should be noted that a cone semi-angle of $\alpha_i = 70.3°$ gives the same area to depth ratio as a triangular Berkovich indenter and is calculated from:

$$\tan\alpha_i = \left(\frac{3\sqrt{3}\tan^2 65.27}{\pi}\right)^{\frac{1}{2}} \tag{3.2.4b}$$

Upon unloading, the contact is elastic and the relationship between the load and the depth of penetration for a cone is given by[7]:

$$P = \frac{2}{\pi}E^* h_e^{\,2}\tan\alpha' \tag{3.2.4c}$$

where α' is now the combined angle of the indenter and the residual impression. The relatively large value of α' means that any effect of radial displacements predicted by the elastic contact equations can be ignored. The normal displacement h of points on the surface beneath the indenter is a function of the radial distance r from the axis of symmetry and is given by:

$$h = \left(\frac{\pi}{2} - \frac{r}{a}\right) a\cot\alpha' \qquad r \le a \tag{3.2.4d}$$

As shown in Fig. 3.4, as the indenter is unloaded, the tip of the indenter ($r = 0$) moves through a distance h_e and the edge of the circle of contact with the specimen surface ($r = a$) moves through a distance h_a.

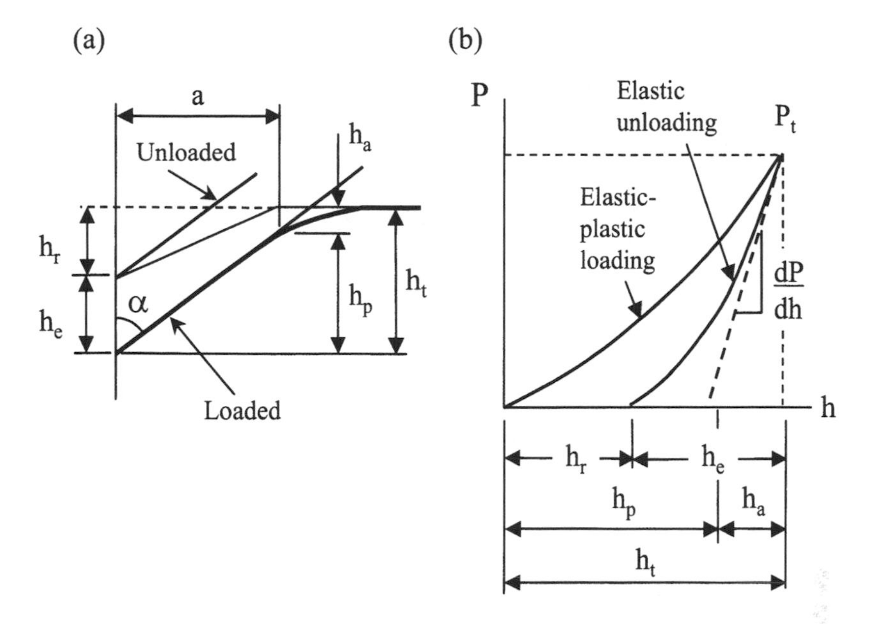

Fig. 3.4 (a) Schematic of indenter and specimen surface at full load and unload for a conical indenter. (b) Load versus displacement for elastic–plastic loading followed by elastic unloading. h_r is the depth of the residual impression, h_t is the depth from the original specimen surface at load P_t, h_e is the elastic displacement during unloading, and h_a is the distance from the edge of the contact to the specimen surface at full load. Upon elastic reloading, the tip of the indenter moves through a distance h_e, and the eventual point of contact with the specimen surface moves through a distance h_a (after reference 1).

Making use of Eq. 3.2.4d, at load P_t the displacements h_e and h_a are thus:

$$h_e = \frac{\pi}{2} a \cot \alpha'$$
$$h_a = \left(\frac{\pi}{2} - 1\right) a \cot \alpha' \tag{3.2.4e}$$

and hence:

$$h_a = \left(\frac{\pi - 2}{\pi}\right) h_e \tag{3.2.4f}$$

and also

$$h_t = h_p + h_a \tag{3.2.4g}$$

We are now in a position to examine the single- and multiple-point methods of analysis for the Berkovich indenter.

3.2.4.1 Multiple-Point Unload Method

The multiple-point unload method uses the slope of the tangent to the initial unloading to determine the quantities of interest. From Eq. 3.2.4c, the slope of the elastic unloading is given by:

$$\frac{dP}{dh} = 2\frac{2E^* \tan\alpha'}{\pi}h_e \qquad (3.2.4.1a)$$

Substituting back into Eq. 3.2.4c, we have:

$$P = \frac{1}{2}\frac{dP}{dh}h_e \qquad (3.2.4.1b)$$

Substituting Eq 3.2.4.1b into 3.2.4f gives:

$$h_a = \left[\frac{2(\pi-2)}{\pi}\right]\frac{P_t}{dP/dh} \qquad (3.2.4.1c)$$

from which h_p can be found from Eq. 3.2.4g and the projected area of contact A from Eq. 3.2.4a and the combined modulus E^* from Eq. 3.2.3.1i. The value of the square-bracketed term depends on the indenter and is given the symbol ε:

$$h_p = h_t - \varepsilon\frac{P_t}{dP/dh} \qquad (3.2.4.1d)$$

For the case of a conical indenter, $\varepsilon \approx 0.72$. For a spherical indenter, $\varepsilon = 0.75$ (see Eq. 3.2.3.1e). Oliver and Pharr[4] found experimental results on standard specimens indicated $\varepsilon = 0.75$ better accounts for the material behavior and they suggest that results from the inevitable tip rounding of a real indenter. There has been some formal justification for this choice of value for ε based upon the nature of the elastic recovery of the specimen material.[8,9,10]

From Eqs. 3.2.4e, and 3.2.4.1a, we have:

$$E^* = \frac{dP}{dh}\frac{1}{2a} = \frac{1}{2}\frac{dP}{dh}\frac{\sqrt{\pi}}{\sqrt{A}} \qquad (3.2.4.1e)$$

So far, the non-symmetrical pyramidal shape of a real Berkovich indenter has not been accounted for. Finite element calculations for indentations formed with flat-ended punches of triangular cross sections[11] yield a correction factor β equal to 1.034 for a three-sided indenter and this is applied to the measured unloading stiffness dP/dh. The value of dPdh to be used in the above equations is determined from the measured experimental value by:

$$\frac{dP}{dh} = \frac{1}{\beta}\frac{dP}{dh}_{\text{measured}} \qquad (3.2.4.1f)$$

3.2.4.2 Single-Point Unload Method

The single-point unload method uses a single unload point rather than a determination of the slope of a fitted line to several unload points. Now, from Eq. 3.2.4g, we have:

$$\begin{aligned} h_t &= h_p + h_a \\ &= h_p + \frac{\pi - 2}{\pi} h_e \end{aligned} \tag{3.2.4.2a}$$

Since $h_e = h_t - h_r$, then

$$h_t = h_p + \frac{\pi - 2}{\pi}(h_t - h_r) \tag{3.2.4.2b}$$

leading to:

$$h_p = h_t - \left[\frac{\pi - 2}{\pi}\right] h_t + \left[\frac{\pi - 2}{\pi}\right] h_r \tag{3.2.4.2c}$$

Now, in a similar manner to the case of a spherical indenter, the depth h_r is found by forming the ratio of elastic displacements for a partial unload P_s:

$$h_r = \frac{h_s (P_t/P_s)^{1/2} - h_t}{(P_t/P_s)^{1/2} - 1} \tag{3.2.4.2d}$$

Equation 3.2.4.2d, substituted into Eq. 3.2.4.2c allows h_p to be determined and hence A from Eq. 3.2.4a and thus hardness H. Elastic modulus can be computed using Eq. 3.2.3.1i.

It is convenient to retain the square-bracketed terms in Eq. 3.2.4.2c as separate entities since this, multiplied by a factor of 2, represents the factor ε in Eq. 3.2.4.1c. The significance of this is that Oliver and Pharr[4] found that to obtain the best correspondence with independently measured mechanical properties, ε in Eq. 3.2.4.1c for a Berkovich indenter should be in fact increased from 0.72 to 0.75 — the same as that for a spherical indenter (see Eq. 3.2.3.1e). By retaining the form of Eq. 3.2.4.2c as shown above, a similar adjustment may be made within the single-point unload method if desired.

It is very important to note that the above treatments rely on some simplifying assumptions, namely: that the unloading is a purely elastic event and that no plasticity occurs; and that the shape of the fully unloaded impression is that of a flat-sided cone (i.e. the sides of the residual impression are straight) or spherical. Finite element analysis shows that the assumption of no reverse plasticity is reasonable, but elastic recovery of the material upon removal of load leads to some significant deviations in the expected shape of the unloading portion of the load-displacement curve. This latter point will be discussed at length in Section 3.2.6.

3.2.5 Knoop Indenter

The analysis methods described above for the Berkovich indenter rely on the conversion of the actual indenter geometry to an equivalent cone. That is, the elastic theory applied to the unloading is for a conical indenter of semi-angle α. The various adjustments that may be made to account for the real indenter geometry have been discussed above. Similar analyses may be applied to the case of Vickers, cube corner, and other indenters (see Table 1.1). However, an interesting issue arises for the case of a Knoop indenter. A Knoop indenter is a four-sided pyramidal indenter with unequal angles such as shown in Fig. 2.4 and where the projected area of contact is given by:

$$A = \frac{d^2}{2}\left[\cot\theta_1 \tan\theta_2\right] \tag{3.2.5a}$$

where $\theta_1 = 86.25°$ and $\theta_2 = 65°$ and where d is the length of the long diagonal. Expressed in terms of the plastic depth h_p, Eq. 3.2.5a becomes:

$$A = 2{h_p}^2 \tan\theta_1 \tan\theta_2 \tag{3.2.5b}$$

Analysis of experimental data obtained with a Knoop indenter on fused silica, using the methods above for an equivalent cone angle of 77.64° show that both the hardness and the modulus are over estimated by about 10% and 50%, respectively. The reason is that there is substantial elastic recovery of the short diagonal of the residual impression compared with negligible elastic recovery of the long axis direction. The long axis of the impression made by a Knoop indenter is approximately seven times larger than the short axis at full load. Upon removal of load, elastic strains stored within the material are relaxed as the specimen material attempts to regain its original shape. Now, since the long axis of the impression made by a Knoop indenter is much greater than the short axis, the restoring forces perpendicular to the long axis (i.e., those resulting from the relaxation of elastic strains on the short axis) have a much longer "moment arm" than those perpendicular to the short axis. In other words, the sides "collapse" inwards as the indenter is withdrawn. A similar effect is demonstrated when breaking an egg by pressing along the long axis as compared with along the short axis.[12] This means that observed elastic recovery in the short axis direction can be substantial compared to that in the long axis direction, especially for materials with a low value of E/Y (or E/H), where elastic recovery is more pronounced (see Fig. 3.5). Other indenters (such as Vickers or Berkovich), while not axial-symmetric, have equal lengths of axes and there is an equal balance of restoring forces on the specimen material during unloading.

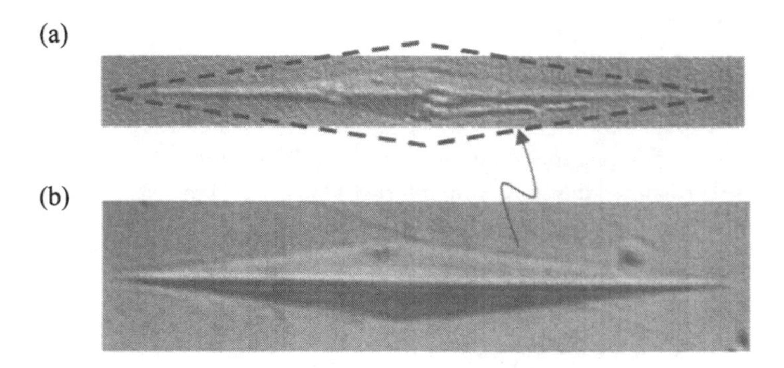

Fig. 3.5 Optical micrographs of residual impression in specimen surface for Knoop indenter at 500 mN load for (a) fused silica and (b) hardened steel. The outline of the impression in (b) is overlaid on the photograph for the fused silica specimen in (a). The measured lengths of the long diagonals are (a) 33 μm and (b) 31 μm and these provide a scale for the figure. The degree of elastic recovery in the fused silica specimen (E/H ≈ 10) is much greater than that for the hardened steel specimen (E/H ≈ 28) (Courtesy CSIRO and after reference 13).

Marshall, Noma, and Evans[14] likened the elastic recovery for a Knoop indenter to that of a cone with major and minor axes. They applied elasticity theory to arrive at an expression for the recovered indentation size in terms of the geometry of the indenter and the ratio H/E:

$$\frac{b'}{d'} = \frac{b}{d} - \alpha \frac{H}{E} \tag{3.2.5c}$$

In Eq. 3.2.5c, α is a geometry factor found from experiments on a wide range of materials to be equal to 0.45. The ratio of the dimension of the short diagonal b to the long diagonal d at full load is given by the indenter geometry and for a Knoop indenter, b/a = 1/7.3. The primed values of d and b are the lengths of the long and short diagonals after removal of load. Since there is observed to be negligible recovery along the long diagonal, it can be said that d' ≈ d. When H is small and E is large (e.g., metals), then b' ≈ b indicating negligible elastic recovery along the short diagonal. When H is large and E is small (e.g., glasses and ceramics), there b' << b is expected.

The elastic analyses described in previous sections rely on Sneddon's solution for a conical indenter in which the depth as a function of load is given by Eq. 3.2.4c. For materials with a low value of E/Y (e.g., metals), it would be expected these analysis methods to give acceptable results even for a Knoop indenter, since the amount of elastic recovery is small. For glass and ceramics, the experimental readings of load and displacement for a Knoop indenter will be affected by the elastic recovery along the short axis dimension and this will not be accommodated by Eq. 3.2.4c. Upon loading, to reach a particular depth of penetration, a higher value of load compared with an equivalent conical indenter would need to be applied to overcome the elastic recovery forces arising from the elastic recovery along the short axis direction. Thus, in an experiment involving a Knoop indenter on say, glass, the depth of penetration at any particular load would be less than for an equivalent conical indenter.

Figure 3.6 shows the region of interest about the short axis of a Knoop indenter. For a load/unload cycle with a Knoop indenter, elastic recovery forces act in addition to those experienced by an equivalent cone owing to the required compression, and subsequent expansion, of the "elastic recovery volume" ABC in Fig. 3.6. This volume goes to zero as the dimension b' approaches b. It is possible to account for this in the analysis methods given here by increasing the effective angle θ_2 so that compression and relaxation of the elastic recovery volume is accommodated. How much should the angle θ_2 be adjusted? Evidently, the adjustment should be a function of E/H and incorporate the results of Eq. 3.2.5c. Rearranging Eq. 3.2.5c and assuming that there is no recovery along the long axis such that d' = d, then:

$$\frac{b'}{b} = 1 - \alpha\frac{d}{b}\frac{H}{E} \tag{3.2.5d}$$

where d/b = 7.11 and $\alpha = 0.45$.[14]

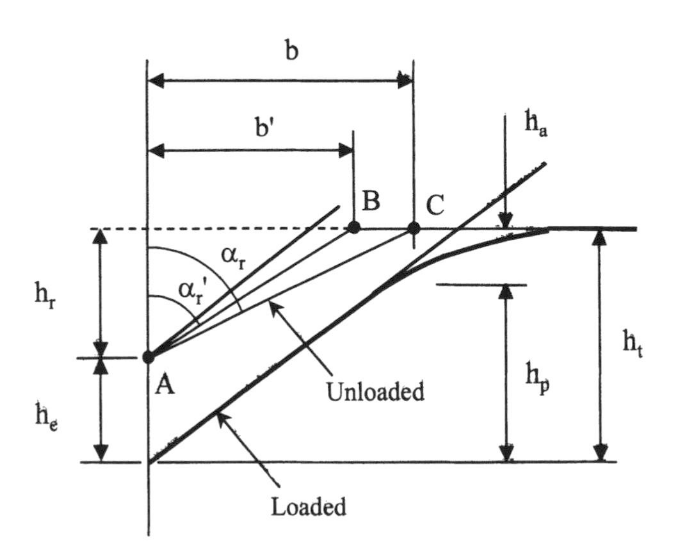

Fig. 3.6 Schematic of the geometry of contact near the recovered impression made by Knoop indenter. When there is elastic recovery along the short axis of the diagonal, the dimension of the short diagonal changes from b to b' and the corresponding change in the angle of the residual impression is from α_r to α_r' (Courtesy CSIRO and after reference 13).

With reference to Fig. 3.6, it can be seen that the angle of the residual impression changes from α_r to α_r' according to:

$$\frac{\tan \alpha_r'}{\tan \alpha_r} = \frac{b'}{b} \tag{3.2.5e}$$

Let us assume that the same fractional change of angle may be attributed to the proposed increase in angle θ_2 of the indenter. The corrected angle θ_2' for the Knoop indenter that accounts for elastic recovery forces is thus:

$$\tan \theta_2' = \left(\frac{b'}{b} \tan \theta_2 \right) \tag{3.2.5f}$$

where b/b' is given by Eq. 3.2.5d. For the purposes of analyses, it can be immediately seen that an initial guess at the ratio of E/H is required for insertion into Eq. 3.2.5d. E and H can then be calculated by the methods described earlier and the ratio E/H adjusted for convergence.

3.2.6 Fitting the Unloading Curve

Finite element analysis and experimental results (see Chapter 5) shows that the underlying assumption of a residual impression with straight sides (for the case of a general conical indenter) is not supported for materials that exhibit significant elastic recovery. The significance of this is that the shape of the unloading curve is represented by a power law whose index depends upon the nature of the specimen material. Pharr and Bolshakov[10] account for this by recognizing that the effective contact angle is changing with depth due to elastic recovery. Their analysis effectively calculates an effective shape of the indenter that results in the same surface displacements when indented into a flat surface. The resulting contact mechanics are described by Sneddon's solution for an axis-symmetric indenter of arbitrary profile[15]. In this scheme, the profile of the indenter is described by a power law:

$$z = Br^n \tag{3.2.6a}$$

where B and n are constants. Using Sneddon's results, Pharr and Bolshakov give the load-displacement response for an indenter with arbitrary profile as:

$$P = \frac{2E^*}{\left(\sqrt{\pi}B\right)^{1/n}} \frac{n}{n+1} \left(\frac{\Gamma(n/2+1/2)}{\Gamma(n/2+1)} \right)^{1/n} h^{1+1/n} \tag{3.2.6b}$$

where Γ is the gamma function‡‡. For the unloading part of the indentation cycle, the load-displacement curve is best represented by a power law function of the form:

$$P = A\left(h - h_r\right)^m \tag{3.2.6c}$$

where A is a constant that depends upon the mechanical properties of the specimen, h_r is the depth of the residual impression and m is the exponent to be determined. It is common practice to fit a 2nd degree polynomial to unloading load-displacement data and for the case of n = 2, this is equivalent to Eq. 3.2.6c since:

$$\begin{aligned} P &= A\left(h - h_r\right)^2 \\ &= Ah^2 - \left(2Ah_r\right)h + \left(Ah_r^{\ 2}\right) \end{aligned} \tag{3.2.6d}$$

which is a quadratic equation in h. For an idealized conical indenter, Eq. 3.2.6d fits the unloading data exactly with the minimum occurring at $h = h_r$. This fact provides justification for the common practice of fitting Eq. 3.2.6d to the initial portion of the unloading data to determine dP/dh at the maximum load. However, when the exponent m in Eq. 3.2.6c is unknown, the fitting procedure is more difficult, it being non-linear in terms of the factors to be determined. If reasonable starting values of the quantities A, h_r and n are given, such as by the use of Eq. 3.2.6d, then an iterative technique[16] (see Appendix 4) can be used which uses a linear approximation approach to the least squares analysis procedure. In Eq. 3.2.6c, the index m is related to n in 3.2.6a by:

$$m = 1 + \frac{1}{n} \tag{3.2.6e}$$

A particular case of interest in n = 1 which yields:

$$P = \frac{2E^*}{\left(\sqrt{\pi}B\right)}\frac{1}{2}(2)h^2 \tag{3.2.6f}$$

and should be compared with Eq. 1.2m.

Now, since by Eq. 3.2.4.1c, the depth of the circle of contact from the specimen free surface is:

$$h_a = \varepsilon\frac{P_t}{dP/dh} \tag{3.2.6g}$$

and that h in Eq. 3.2.6b is found from $h_t - h_a$, then the parameter ε can be expressed:[8,10]

‡‡ $\Gamma(1) = 1$, $\Gamma(3/2) = 1/2\pi^{1/2}$, $\Gamma(2) = 1$, $\Gamma(1/2) = \pi^{1/2}$

$$\varepsilon = \left(1 - \frac{2\Gamma(1/2 + n/2)}{n\sqrt{\pi}\Gamma(n/2)}\right)\frac{1+n}{n} \tag{3.2.6h}$$

which for the case n = 1 (m = 2) gives 0.72. The commonly used value of ε = 0.75 for a Berkovich indenter is thus a consequence of the curved shape of the residual impression which is quantified by the concept of an effective indenter shape as described by Eq. 3.2.6a. Thus, the index m be determined directly from fitting the experimental data, ε calculated from Eq. 5.6c, and the multiple-point unloading method then applied in the normal manner. A method for fitting Eq. 3.2.6c to experimental data is given in Appendix 4. While this method provides a more accurate approach to analyzing the unloading data, in practice, a 2nd order polynomial fit applied to the *initial* unloading data (upper 30% or so) results in only a small error in the calculated quantities.

It should be noted that the value of modulus calculated using Eqs. 3.2.3.1i and 3.2.4.1e depends upon the stiffness, dP/dh *and* the area of contact at any given load. This means that the shape of the unloading curve is not solely dependent upon E^* but also depends upon the hardness of the material. For two materials with the same value of E^*, but different hardness, for a smaller value of H, the slope dP/dh will be increased to compensate for the increase in h_p.

It is of interest to note that the ISO draft standard[17] for instrumented indentation testing specifies that the upper 80% of the unloading data be used in the fitting procedure which can take the form of a linear extrapolation or power law fitting.

3.2.7 Energy Methods

Hardness from nanoindentation measurements can be measured by a consideration of the energies involved in the loading and unloading processes. In general, the indentation process consists of an elastic–plastic loading followed by an elastic unloading in which the load P and depth of penetration h are related by the general expression:

$$P = Ah^m \tag{3.2.7a}$$

where the constant A depends on the nature of the contact: elastic, elastic plastic, or purely plastic. The index m = 2 for a conical indenter. From the arguments presented in Chapter 5, it is shown that for an elastic–plastic contact with a conical indenter of semi-angle α, the value of A during loading can be expressed as:

$$A_p = \left[\frac{1}{\sqrt{\pi H \tan^2\alpha}} + \left[\frac{2(\pi - 2)}{\pi}\right]\sqrt{\frac{\pi}{4}}\frac{\sqrt{H}}{E^*}\right]^{-2} \tag{3.2.7b}$$

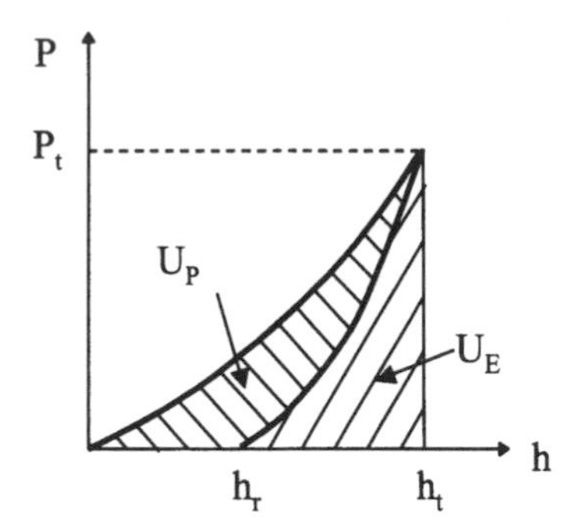

Fig. 3.7 The net area U_p enclosed by the load-displacement curve represents the energy lost due to plastic deformation and elastic strains for residual stresses in the specimen material.

For the elastic unloading, from Eq. 1.2m, we have:

$$A_e = \frac{2}{\pi} E^* \tan\alpha \tag{3.2.7c}$$

Now, application of a load P to the indenter and the resulting displacement represents work done on the system and is manifested as both heat (via plastic flow) and elastic strains within the specimen. During unloading, work is done by the system as the material partially recovers in an elastic manner. The proportion of energy returned to the total energy supplied is dependent upon the ratio E^*/H of the specimen material and appears to be relatively independent of the shape of the indenter.[18]

The net area enclosed by the load-displacement response shown in Fig. 3.7 represents the energy U_p lost as heat in plastic deformation as well as stored elastic energy (from residual stresses) within the specimen material. At the condition of maximum load, $P = P_t$ we have:

$$P = A_p h_t^2 = A_e (h_t - h_r)^2 \tag{3.2.7d}$$

The net work of indentation is thus the area enclosed by the loading and unloading curves and be calculated by integrating P with respect to h:

$$\begin{aligned} U_p &= \int_0^{h_t} A_p h^2 dh - \int_{h_r}^{h_t} A_e (h - h_r)^2 dh \\ &= (A_p - A_e)\frac{h_t^3}{3} \end{aligned} \tag{3.2.7e}$$

In one extreme of material behavior, a rigid–plastic material, we can set E^* to infinity in Eq. 3.2.7b. In this case, there is no elastic recovery and $h_t = h_r$ in Eq. 3.2.7e and thus:

$$A_p = \pi H \tan^2 \alpha \tag{3.2.7f}$$

and

$$U_p = A_p \frac{h_t^{\,3}}{3} \tag{3.2.7g}$$

Substituting Eq. 3.2.7f into Eq. 3.2.7g, and using the fact that the volume of a cone of radius a and height h_t is given by:

$$V = \frac{\pi}{3} a^2 h_t = \frac{\pi}{3} a^3 \cot \alpha \tag{3.2.7h}$$

and $\tan \alpha = a/h_t$, it can be shown that:

$$H = \frac{U_p}{V} \tag{3.2.7i}$$

where hardness is thus seen to be equivalent to the work required to produce a unit volume of indentation. This hardness is called the true hardness since it is a true measure of resistance to plastic deformation.

Also, using Eqs. 3.2.7a and 3.2.7f, and substituting into Eq. 3.2.7g, we obtain:[19]

$$U_p = \left(\frac{1}{3} \sqrt{\frac{.\,1}{\pi \tan^2 \alpha}} \right) \frac{1}{\sqrt{H}} P^{3/2} \tag{3.2.7j}$$

The slope of the line of best fit of experimentally derived values of U_p plotted against $P^{3/2}$ provide a measure of the true hardness H.

Eq. 3.2.7j also applies to the more general case of elastic–plastic contact where U_p is the net area under the load-displacement curve as computed using Eq. 3.2.7e. In this case, Eq. 3.2.7j gives the true hardness is that which would be attributed to the specimen material if it had deformed as a rigid–plastic solid. It is a measure of the irrecoverable energy dissipative (heat) and elastic strain (from the residual stress field) required to create a unit volume of indentation.

In an elastic–plastic contact, elastic recovery of the specimen occurs which cases the residual depth of impression to the less than the maximum depth. In this case, the volume of the indentation (i.e., the net volume swept out by the indenter) is the volume of the residual impression and is calculated from:

$$V_r = \frac{\pi}{3} a^2 h_r \tag{3.2.7k}$$

where it is assumed that the radius of the circle of contact is the same at full load and full unload. Since the volume of the residual impression represents the net indentation volume, the hardness is computed from:

$$H = \frac{U_p}{V_r} \tag{3.2.7l}$$

This value of hardness is called the apparent hardness. It is the hardness usually associated with the analysis of the unloading curve and is more correctly termed the limiting value of mean contact pressure. For an ideal rigid–plastic material response, the apparent hardness is equivalent to the true hardness. For an elastic–plastic response, the apparent hardness is less than the true hardness by an amount dependent on the ratio E^*/H. The apparent hardness is less than the true hardness because it measures of the resistance of the material to elastic and plastic deformations, not just the resistance of the material to plastic deformation alone.

The energy balance associated with the indentation can be equivalently considered from the point of view of an increment of indenter displacement. The total work done on the indenter is related to an incremental change in volume of material displaced by the indenter:

$$P\delta h = \frac{P}{A}\delta V \tag{3.2.7m}$$

where P/A is the mean contact pressure equivalent to the apparent hardness H. Additional terms may be shown on the right hand side of Eq. 3.2.7m that account for the increase in surface area of the impression (i.e. surface energy) and interfacial friction.[20,21]

It should be noted that the volume of material referred to in the above discussion is that displaced by the indenter, not the volume of the plastic zone. The volume of the plastic zone simply indicates the extent over which the relevant plastic failure criterion is satisfied. For a geometrically similar indentation there is a correspondence between the two since the volume of the plastic zone and that displaced by the indenter scale in the same proportion.

3.2.8 Dynamic Methods

Measurement of elastic modulus and hardness involves the measurement of the contact stiffness, dP/dh, at the loading point (see Eq. 3.2.3.1i). In a typical indentation test involving an oscillatory motion, a small AC modulated force p is applied with a frequency ω and amplitude p_o[22]:

$$p = p_o e^{i\omega t} \tag{3.2.8a}$$

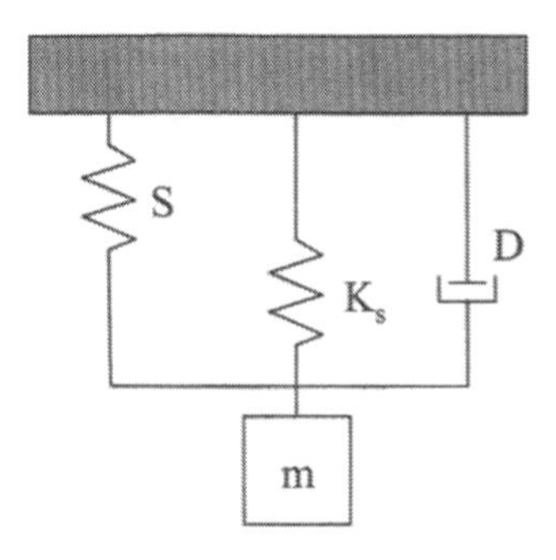

Fig. 3.8 Dynamic mechanical model of a nanoindentation instrument.

The resulting displacement h will have the same frequency of oscillation but may have a phase difference ϕ leading to:

$$h = h_o e^{i(\omega t + \phi)} \tag{3.2.8b}$$

These displacements will be affected by the dynamics of the instrument as well as the indenter-specimen interaction. The magnitude of the stiffness of the contact is found by summing the stiffness and damping terms shown in Fig. 3.8.

$$\left|\frac{p_o}{h_o}\right| = \sqrt{\left(S + K_s - m\omega^2\right)^2 + \omega^2 D^2} \tag{3.2.8c}$$

The phase difference ϕ between the load and depth terms is given by:

$$\tan\phi = \frac{\omega D}{S + K_s - m\omega^2} \tag{3.2.8d}$$

In Eqs. 3.2.8c and 3.2.8d, K_s is the stiffness of the indenter shaft support springs, p_o is the magnitude of the oscillatory load, D is a damping coefficient, h_o and p_o are the magnitudes of the displacement and force oscillations, ω is the frequency of the oscillation, m is the mass of the components, and S is the contact stiffness dP/dh. Various methods of analysis allow the contact area to be determined and hence the hardness and modulus from Eq. 3.2.1c and 3.2.3d. The technique is popularly used to measure the contact stiffness continuously as the depth of penetration increases by superimposing an oscillatory motion onto the conventional load-displacement response.

Using a similar approach, Lorenz, Fränzel, Einax, Grau, and Berg[23] have demonstrated a method of determining the modulus of elastic materials (low values of E/H) using a nanoindentation technique whereby the contact force is modulated by an oscillatory motion of the specimen. The contact stiffness is obtained from the resonant response of the system. Other systems are designed to measure the viscoelastic properties of the specimen by measuring the phase difference in the force and depth signals resulting from the oscillatory motion. Further developments of this mode of analysis are given in Chapter 7.

3.2.9 Other Methods of Analysis

The analysis methods given in the previous sections are not the only means of determining material properties from indentation test data, but are presented as being the most common, and all utilize the elastic unloading response of a depth-sensing indentation test as the basis of the measurement. Other methods generally focus on the shape of the loading and unloading curves either separately or together.

Joslin and Oliver[24] proposed that a new parameter that characterizes the resistance to plastic penetration, H/E^2, could be obtained from measurements of the contact stiffness (dP/dh). The method has the advantage of not requiring the precise shape of the indenter to be determined and is less affected by surface roughness.

Hainsworth, Chandler, and Page[25] showed that the loading curve of a specimen could be described using a linear relationship between the load P and the square of the displacement:

$$P = Ah^2 \tag{3.2.9a}$$

Superimposing the displacements arising from both elastic and plastic deformation, the constant of proportionality A was found to be:

$$A = E\left(\phi_m\sqrt{\frac{E}{H}} + \varphi_m\sqrt{\frac{H}{E}}\right)^{-2} \tag{3.2.9b}$$

where ϕ_m and φ_m are constants that evaluate to 0.194 and 0.930 for a Berkovich indenter, respectively. Equations 3.2.9a and 3.2.9b have been shown to provide a good fit to experimental data for specimens with a wide range of modulus and hardness values. It should be also noted that Eqs. 3.2.9a and 3.2.9b, with a little rearrangement, are consistent with Eq. 2.4c.

It is possible to determine mechanical properties of interest in a "forward" direction rather than the "reverse" direction as shown above. For example, using material property parameters as inputs, the expected load-displacement response can be predicted and compared with that obtained experimentally. Alternately, the expected P vs h^2 relationship for an ideally sharp indenter may be used to determine E and H by fitting this function to the experimental loading curve. Venkatesh, Van Vliet, Giannakopolous, and Suresh[26] computed the load-displacement curves for a wide variety of material properties and varied these by ±5% to determine the sensitivity of the method to small variations in the inputs. They found that the least sensitive quantity was the plastic to elastic work ratio. In the reverse analysis, using the load-displacement curve as an input to determine mechanical properties, it was found that small variations in the unloading slope dP/dh and curvature of the loading response had little effect on the final results, but a strong sensitivity was observed for small changes in the elastic to plastic work ratio.

Gerberich, Yu, Kramer, Strojny, Bahr, Lilleodden, and Nelson[27] summarize the various methods of analysis and propose another scheme based upon the elastic and plastic deformations associated with contact using a spherical indenter. In an elastic contact with a spherical indenter, the depth of the circle of contact is exactly half that of the total depth of penetration (see Eq. 3.2.3c). These authors assumed (with justification) that when plastic deformation occurs, the depth of the circle of contact measured from the specimen free surface (h_a) decreases in proportion to the elastic displacement h_e and, at the same time, decreases in proportion to the elastic displacement at large indentation depths according to:

$$h_a = \frac{h_e}{2}\left(\frac{h_{e\max}}{h_{t\max}}\right) \tag{3.2.9c}$$

The terms with the "max" subscript are evaluated at some large indentation load and Eq. 3.2.9c is then applied to indentations made at smaller loads as desired. For a completely elastic contact, $h_{emax} = h_{tmax}$ and Eq. 3.2.9a reverts to the Hertzian contact condition of $h_a = h_e/2$. Once h_a is known, then the plastic depth h_p and hence the area of the contact can be determined from which follows E^* and H for the specimen material.

More recently, Page, Pharr, Hay, Oliver, Lucas, Herbert, and Riester[28] have demonstrated that the parameter given by the load divided by the contact stiffness (dP/dh) squared provides material property information from load-displacement curves:

$$\frac{P}{S^2} = \frac{\pi H}{4E^{*2}} \tag{3.2.9d}$$

One major advantage of using this technique is that it does not require a detailed knowledge of the indenter geometry (i.e., the tip shape). An interesting consequence of this is that the technique is not so sensitive to changes in hardness but is sensitive to changes in modulus E^* while the converse appears to be the case for estimations of properties based upon P vs h^2 relationships.

Equation 3.2.9d bears a close relationship to an earlier theoretical treatment by Stone,[29] and Yoder, Stone, Hoffman, and Lin[30] who propose a theoretical model which proposes that the parameter:

$$\frac{\sqrt{P}}{S} = \frac{\sqrt{H}}{E^*} + C_f P \tag{3.2.9e}$$

can be used to obtain a modulus and hardness for thin film specimens where C_f is the compliance of the load frame. These authors concluded that for layered specimens, the conventional methods of analysis are unreliable due to the influence of the substrate, and the values obtained from Eq. 3.2.9e are also somewhat unreliable, but the two techniques, used in conjunction with each other, permit some measure of verification and estimation of the confidence in the values obtained. The appropriateness of Eqs. 3.2.9d and 3.2.9e are related to the theoretical relationships given by Eqs. 3.2.8e and 3.2.3.1i.

Oliver[31], and more recently, Malzbender and de With[32] attempted to correlate the ratio of the loading S_L and unloading slope S_U of the load-displacement curves with theoretical models for the deformation for determining the modulus and hardness. Oliver finds:

$$\frac{S_U}{S_L} = \frac{E^*}{H\sqrt{\pi C}} + \frac{\varepsilon}{2} \tag{3.2.9f}$$

whereupon expressions for the modulus and hardness are given as:

$$\begin{aligned} E^* &= \sqrt{\frac{\pi}{C}} \frac{1}{2P\beta}\left(\frac{S_U{}^2 S_L}{2S_U - \varepsilon S_L}\right)^2 \\ H &= \frac{1}{CP}\left(\frac{2S_U - \varepsilon S_L}{S_U S_L}\right)^{-2} \end{aligned} \tag{3.2.9g}$$

In Eqs. 3.2.9f and 3.2.9g, C is a constant dependent on the shape of the indenter (24.56 for a Berkovich indenter). β is the shape factor (see Eq. 3.2.4.1f) and ε is the square-bracketed term in Eq. 3.2.4.1c.

For a Berkovich indenter, Malzbender and de With obtain a similar expression to the above and show that:

$$\frac{S_L}{S_U} = \frac{\pi \tan\alpha}{\beta} \frac{H}{E^*} \frac{h_p}{h_t} \tag{3.2.9h}$$

from which the ratio H/E^* is obtained. Using Eq. 3.2.3.1i, an expression for E^* and H can be obtained. For E^*, the expression is given as:

$$E^* = \frac{S_L S_U{}^2}{2f(n)S_U - \varepsilon S_L} \frac{1}{2P\beta \tan\alpha} \tag{3.2.9i}$$

where f(n) is a correction factor that accounts for pile-up and sink-in.

Recently, Sakai and Nakano[33] undertook an extensive study on a wide range of engineering materials and focused their attention on the ratio of the maximum depth of penetration to the plastic or contact depth $\gamma = h_t/h_p$. As a result of many experiments, these authors determined the following empirical relationship for evaluating γ in term of the ratio of the depth of the residual impression to the maximum depth $\xi_r = h_r/h_t$.

$$\gamma = \frac{\pi}{2}\left(1 - 0.43\sqrt{\xi_r}\right) \tag{3.2.9j}$$

The multiple-point unload method of Oliver and Pharr is expressed in terms of γ as:

$$\frac{1}{\gamma} = 1 - \varepsilon \frac{P_t}{Sh_t} \tag{3.2.9k}$$

where S is the unloading stiffness dP/dh. Comparison of the empirically derived Eq. 3.2.9j with Eq. 3.2.9k (with $\varepsilon = 0.72$) shows a 10 to 20% difference depending on the material. Sakai and Nakano define a measure of hardness as that being described by the loading parameter k_1:

$$k_1 = \frac{H}{\left[1 + \sqrt{\frac{2H}{E^* \tan\theta}}\right]^2} \frac{g}{\gamma^2} \tag{3.2.9l}$$

where $P = k_1 h^2$ and k_1 is comparable to A in Eq. 3.2.9a. In Eq. 3.2.9k, g is a geometry term dependent on the indenter as shown in the second column of Table 1.1 (e.g. g = 24.5 for a Berkovich indenter) and θ is the face angle for a pyramidal indenter (e.g. 65.27 for a Berkovich indenter). Elastic modulus is found from Eq. 3.2.1c with A being given in terms of γ:

$$A = \frac{g}{\gamma^2} h_t^{\,2} \tag{3.2.9m}$$

where it will be remembered that $\gamma = h_t/h_p$. The consequence of this work is that the hardness H and modulus E^* are both determined in terms of the parameter γ which has been empirically determined according to Eq. 3.2.9j. The value of this is that γ appears to reflect the true elastic–plastic deformation of a wide variety of materials independent of the indenter shape.

Zeng and Chiu[34], and Zeng and Shen[35] report an empirical approach to the analysis of unloading curves based upon a large amount of experimental data and finite element analysis. In this scheme, the unloading curve is written as a combination of two extremes of a perfectly elastic contact and a perfectly elastic–plastic contact. The unloading response is given by:

$$P = (1-\theta) f(\nu) E h2 + 2\theta\left(\frac{24.56}{\pi}\right)^{\frac{1}{2}} \frac{E}{1-\nu^2} h_o (h - h_o) \tag{3.2.9n}$$

where θ is a weighting parameter for the elastic and elastic–plastic nature of the contact, $f(\nu)$ is a Poisson's ratio effect, and h_o is a constant that serves as a fitting parameter whose physical significance is the plastic depth h_p for a perfectly elastic–plastic contact. An iterative procedure is then used to fit experimental data to

Eq. 3.2.9n to determine E and subsequently H. The strain-hardening index is shown to be equal to the fraction θ.

The above techniques are based on instrumented, or depth-sensing indentation experiments. However, nanoindentation, at least for hardness measurements, can also be used with post-facto imaging using high resolution scanning force microscopy (SFM). In some respects, this procedure has some similarities to traditional hardness testing at larger length scales but there are some important differences. Randall and Julia-Schmutz[36] and Randall[37] have used SFM to obtain surface images and profiles of indentations in different metals deposited on to silicon substrates (see Chapter 12). For cases such as these, with a relatively soft coating on a hard substrate, piling-up of the specimen material around the indenter is particularly severe and conventional methods of analysis using depth-sensing indentation will over-estimate the hardness of the material. Unlike optical imaging in large-scale hardness tests, SFM offers quantitative measurements of the surface profile of the residual impression, including the surrounding specimen surface. Scanning force microscopy appears to be the only method to give accurate dimensional information about the indentation and hence a reliable and accurate measurement of hardness. The measurement technique can be semi-automated. Quantities that can be measured are surface roughness, piling-up and sinking-in, the true area of contact, volume of material displaced and the indenter tip shape.

References

1. A.C. Fischer-Cripps, *Introduction to Contact Mechanics*, Springer-Verlag, New York, 2000.
2. G.M. Pharr, W.C. Oliver, and F.R. Brotzen, "On the generality of the relationship among contact stiffness, contact area, and the elastic modulus during indentation," J. Mater. Res. 7 3, 1992, pp. 613–617.
3. M.F. Doerner and W.D. Nix, "A method for interpreting the data from depth-sensing indentation instruments," J. Mater. Res. 1 4, 1986, pp. 601–609.
4. W.C. Oliver and G.M. Pharr, "An improved technique for determining hardness and elastic modulus using load and displacement sensing indentation experiments," J. Mater. Res. 7 4, 1992, pp. 1564–1583.
5. J.S. Field and M.V. Swain, "A simple predictive model for spherical indentation," J. Mater. Res. 8 2, 1993, pp. 297–306.
6. E.S. Berkovich, "Three-faceted diamond pyramid for micro-hardness testing," Ind. Diamond Rev. 11 127, 1951, pp. 129–133.
7. I.N. Sneddon, "Boussinesq's problem for a rigid cone," Proc. Cambridge Philos. Soc. 44, 1948, pp. 492–507.
8. J.Woirgard and J-C. Dargenton, "An alternative method for penetration depth determination in nanoindentation measurements," J. Mater. Res. 12 9, 1997, pp. 2455–2458.

9. T. Sawa and K. Tanaka, "Simplified method for analyzing nanoindentation data and evaluating performance of nanoindentation instruments," J. Mater. Res. 16 11, 2001, pp. 3084–3096.
10. A. Bolshakov and G.M. Pharr, "Understanding nanoindentation unloading curves," J. Mater. Res. 17 10, 2002, pp. 2660–2671.
11. R.B. King, "Elastic analysis of some punch problems for a layered medium," Int. J. Solids Structures, 23 12, 1987, pp. 1657–1664.
12. D.B. Marshall and B.R. Lawn, "Indentation of Brittle Materials," *Microindentation Techniques in Materials Science and Engineering*, ASTM STP 889, P.J. Blau and B.R. Lawn, Eds. American Society for Testing and Materials, Philadelphia, 1986, pp. 26–46.
13. L. Riester, T.J. Bell, and A.C. Fischer-Cripps, "Analysis of depth-sensing indentation tests with a Knoop indenter," J. Mater. Res. 16 6, 2001, pp. 1660–1667.
14. D.B. Marshall, T. Noma, and A.G. Evans, "A simple method for determining elastic-modulus-to-hardness ratios using Knoop indentation measurements," J. Am. Ceram. Soc. 65, 1980, pp. C175–C176.
15. I.N. Sneddon, "The relation between load and penetration in the axisymmetric Boussinesq problem for a punch of arbitrary profile", Int. J. Eng. Sci. 3, 1965, pp. 47–57.
16. A.C. Fischer-Cripps, "Elastic recovery and reloading of hardness impressions with a conical indenter," Mat. Res. Symp. Proc. 750, 2003, pp. Y6.9.1–Y6.9.6.
17. ISO 14577, "Metallic materials — Instrumented indentation test for hardness and materials parameters." ISO Central Secretariat, 1 rue de Varembé, 1211 Geneva 20 Switzerland.
18. M.Kh. Shorshorov, S.I. Bulychev, and V.O. Alekhin, "Work of plastic and elastic deformation during indenter indentation," Sov. Phys. Dokl. 26 8, 1981, pp. 769–771.
19. M. Sakai, "Energy principle of the indentation-induced inelastic surface deformation and hardness of brittle materials," Acta. Metal. Mater. 41 6, 1993, pp. 1751–1758.
20. J.B. Quinn and G.D. Quinn, "Indentation brittleness: a fresh approach," J. Mat. Sci. 32 1997, pp. 4331–4346.
21. T.-Y. Zhang and W.-H. Xu, "Surface effects on nanoindentation," J. Mater. Res. 17 7, 2002, pp. 1715–1720.
22. B.N. Lucas, W.C. Oliver, and J.E. Swindeman, "The dynamics of frequency specific, depth sensing indentation testing," Mat. Res. Soc. Symp. Proc. 522, 1998, pp. 3–14.
23. D. Lorenz, W. Fränzel, M. Einax, P. Grau, and G. Berg, "Determination of the elastic properties of glasses and polymers exploiting the resonant characteristic of depth-sensing indentation tests," J. Mater. Res. 16 6, 2001, pp. 1776–1783.
24. D.L. Joslin and W.C. Oliver, "A new method for analyzing data from continuous depth-sensing microindentation tests," J. Mater. Res. 5 1, 1990, pp. 123–126.
25. S.V. Hainsworth, H.W. Chandler, and T.F. Page, "Analysis of nanoindentation load-displacement loading curves," J. Mater. Res. 11 8, 1996, pp. 1987–1995.
26. T.A. Venkatesh, K.J. Van Vliet, A.E. Giannakopolous, and S. Suresh, "Determination of elasto-plastic properties by instrumented sharp indentation: Guidelines for property extraction," Scripta Mater. 42, 2000, pp. 833–839.

27. W.W. Gerberich, W. Yu, D. Kramer, A. Strojny, D. Bahr, E. Lilleodden, and J. Nelson, "Elastic loading and elastoplastic unloading from nanometer level indentations for modulus determinations," J. Mater. Res. 13 2, 1998, pp. 421–439.
28. T.F. Page, G.M. Pharr, J.C. Hay, W.C. Oliver, B.N. Lucas, E. Herbert, and L. Riester, "Nanoindentation characterization of coated systems: P:S2 – a new approach using the continuous stiffness technique," Mat. Res. Symp. Proc. 522, 1998, pp. 53–64.
29. D.S. Stone, "Elastic rebound between an indenter and a layered specimen: Part 1. Model," J. Mater. Res. 13 11, 1998, pp. 3207–3213.
30. K.B. Yoder, D.S. Stone, R.A. Hoffman, and J.C. Lin, "Elastic rebound between an indenter and a layered specimen: Part II. Using contact stiffness to help ensure reliability of nanoindentation measurements," J. Mater. Res. 13 11, 1998, pp. 3214–3220.
31. W.C. Oliver, "Alternative technique for analyzing instrumented indentation data," J. Mater. Res. 16 11, 2001, pp. 3202–3206.
32. J. Malzbender and G. de With, "Indentation load-displacement curve, plastic deformation, and energy," J. Mater. Res. 17 2, 2002, pp. 502–511.
33. M. Sakai and Y, Nakano, "Elastoplastic load-depth hysteresis in pyramidal indentation," J. Mater. Res. 17 8, 2002, pp. 2161–2173.
34. K. Zeng and C.-h Chiu, "An analysis of load-penetration curves from instrumented indentation," Acta Mater. 49, 2001, pp. 3539–3551.
35. K. Zeng and L. Shen, "A new analysis of nanoindentation load-displacement curves," Phil. Mag. A 82 10, 2002, pp. 2223–2229.
36. N.X. Randall and C. Julia-Schmutz, "Evolution of contact area and pile-up during the nanoindentation of soft coatings on hard substrates," Mat. Res. Symp. Proc. Vol. 522, 1998, pp. 21–26.
37. N.X. Randall, "Direct measurement of residual contact area and volume during the nanoindentation of coated materials as an alternative method of calculating hardness," Phil. Mag. A 82, 10, 2002, pp. 1883–1892.

Chapter 4
Factors Affecting Nanoindentation Test Data

4.1 Introduction

In conventional indentation tests, the area of contact between the indenter and the specimen at maximum load is usually calculated from the diameter or size of the residual impression after the load has been removed. The size of the residual impression is usually considered to be identical to the contact area at full load, although the depth of penetration may of course be significantly reduced by elastic recovery. Direct imaging of residual impressions made in the submicron regime are usually only possible using inconvenient means and, for this reason, it is usual to measure the load and depth of penetration directly during loading and unloading of the indenter. These measurements are then used to determine the projected area of contact for the purpose of calculating hardness and elastic modulus. In practice, various errors are associated with this procedure. The most serious of these errors manifests themselves as offsets to the depth measurements. Others arise from environmental changes during the test and the non-ideal shape of the indenter. In addition to the above, there are a number of materials related issues that also affect the validity of the results. The most serious of these are an indentation size effect and the phenomenon of piling-up and sinking-in. The sensitivity of nanoindentation tests to these phenomena and others is a subject of continuing research.[1] In this chapter, some of the most commonly encountered sources of error and methods of accounting for them are reviewed.

4.2 Thermal Drift

There are two types of drift behavior that might be observed in nanoindentation testing. The first is creep within the specimen material as a result of plastic flow. Creep may manifest itself most clearly when the load is held constant, and the depth readings increase as the indenter sinks into the specimen. Another reason for an observed change in depth with constant load that is virtually indistinguishable from specimen creep is a change in dimensions of the instrument due to thermal expansion or contraction of the apparatus. This change in depth imposes a thermal drift error onto the real depth of penetration readings. If the rate of change of depth reading with time is measured for a constant value of load at some point during an indentation test, then the thermal drift rate can be computed and the depth readings taken throughout the test adjusted accordingly.

Feng and Ngan[2] undertook a detailed study on the effect of thermal drift to determine under what conditions it would have a negligible effect on the computed value of elastic modulus. They found that if t_h is the total time from the start of the test to the beginning of the unloading, then it is shown that the thermal drift has minimum effect on the computer elastic modulus (from an analysis of the unloading curve) if

$$t_h \approx \frac{S}{|\dot{P}|} h_p \tag{4.2a}$$

where S is the contact stiffness and $\dot{P}$ is the unloading rate.

One particular source of thermal drift that is generally not considered in nanoindentation testing is that due to the generation of heat within the plastic zone of the indentation. Calculations[3] show that the temperature rise within the specimen material may be quite substantial (≈100 °C) but the volumes of material are so small that any change in linear dimension of the specimen is less than 0.1% of the overall penetration depth and can be safely ignored. However, localized high temperatures within the specimen may affect the viscosity and hardness of the material[4] and, in some cases, may be worthy of further consideration.

To correct for thermal drift, some nanoindentation instruments allow for a hold series of data points to be accumulated at either maximum load or at the end of the unloading from maximum load. For the purposes of calculating thermal drift, the hold period data at the final unload increment should be used, since this is done at a low value of load where creep within the material is less likely to occur, especially for the case of a spherical indenter. When creep properties of the specimen material are of interest, then hold at maximum load is more appropriate. A linear regression to the load-displacement response within the hold period can be used to obtain a thermal drift rate. The thermal drift rate is then applied to all the depth readings according to the time at which they were logged during the actual test. A good test of the effectiveness of the correction is that the thermal drift rate, when applied to the thermal drift data, should collapse that data into a single point, or very close thereto.

4.3 Initial Penetration Depth

In a nanoindentation test, the indenter displacement is ideally measured from the level of the specimen free surface. However, in practice, the indenter must first make contact with the specimen surface before the displacement measurements can be taken. That is, in practice, it is necessary to make actual contact with the specimen surface to establish a datum for the test displacement measurements. This initial contact depth is usually made to be as small as possible, and is often set using the smallest obtainable force of the instrument. An initial contact force on the order of 1 μN is usually achievable.

However, no matter how small the initial contact force is made, there is a corresponding penetration of the indenter beneath the undisturbed specimen free surface as shown schematically in Fig. 4.1. Thus, all subsequent displacement measurements taken from this datum will be in error by this small initial penetration depth. The initial penetration depth, h_i, has to be added to all displacement measurements, h, to correct for this initial penetration.

For contact with a spherical indenter, it may be assumed that the first few loading points result in a purely elastic deformation of the specimen. It is then possible to model these initial data points using the Hertz equations. The Hertz equations predict that the relationship between load and penetration for an elastic response is of the form:

$$h \propto P^m \tag{4.3a}$$

where m = 2/3 for a spherical indenter, m = 1 for a cylindrical flat punch indenter, and m = 1/2 for a conical indenter.

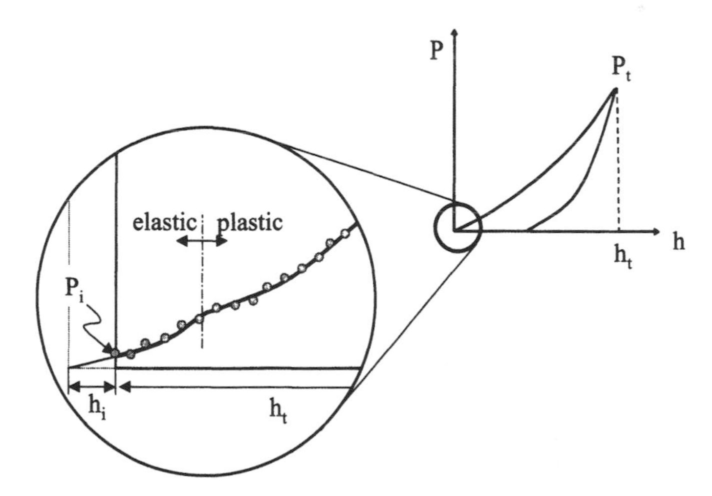

Fig. 4.1 Schematic of the effect of initial penetration depth on load-displacement data for a depth-sensing indentation test. The initial contact load P_i results in an initial penetration depth h_i. Depth readings h at loads P must be corrected for h_i.

At the initial contact load P_i, there is an initial depth h_i. During this initial loading, where the response is elastic, the instrument measures P and h where the penetration $h' = h + h_i$ at each load. Thus,

$$h + h_i \propto P^m \tag{4.3b}$$

and therefore

$$\begin{aligned} h &= kP^m - h_i \\ &= kP^m - kP_i^m \end{aligned} \tag{4.3c}$$

where k is a constant whose value depends upon the shape of the indenter. For an initial elastic response, say for the first five or so data points in a typical test, a series of values for h and P is obtained, and also a value for P_i. The terms m and k are the unknowns. Once values for m and k are found, the initial penetration h_i can be found from:

$$h_i = kP_i^m \tag{4.3d}$$

From Eq. 4.3c, a plot of h vs $P^m - P_i^m$ should be linear with a slope k. The easiest way to adjust the variables m and k for a linear response is to plot the logarithm of both sides to obtain a slope equal to unity. Thus:

$$\log h = \log k + A \log\left(P^m - P_i^m\right) \tag{4.3e}$$

where A = 1. A plot of log h vs log ($P^m - P_i^m$) should have a slope A = 1 if m and k are chosen correctly. Note that the first data point of the experimental data with h = 0 and $P = P_i$ is not included in the fitting procedure owing to the impossibility of taking the log of zero.

As mentioned previously, the initial contact with the specimen surface is usually made using the smallest possible indenter load of the instrument. Under these conditions, the measuring instrument is typically operating at the limit of its resolution. Thus, the actual load applied to the indenter may be substantially different from the user-specified value. The difference between the actual and specified loads decreases with increasing load as the instrument moves into a more stable operating regime. To minimize the instrumentation errors at the minimum load P_i, the slope A of a plot of log h_t vs log ($P^m - P_i^m$) is optimized by first adjusting P_i with m set to m = 2/3 until the slope A in Eq. 4.3e is as close to possible to unity. Although the value of h_i has not yet been estimated, this initial fitting reduces the effect of any errors in the recorded value of P_i.

The adjusted initial contact force P_i and measured values of P and h are then plotted according to Eq. 4.3e and the value of m varied until the slope A is as close to unity as possible. The resulting intercept provides a value for k. A value for h_i can then be determined from Eq. 4.3d. The corrected depth h' is thus:

$$h' = h + h_i \tag{4.3f}$$

Care must be taken with the choice of the number of initial data points to be used in the analysis. It is assumed that the material response is elastic and thus one must remain within the elastic regime otherwise the fitting will be influenced by data which reflect plastic deformation, in which case Eq. 4.3a does not apply.

Alternately, a nonlinear least squares fitting procedure (see Appendix 4) can be used to fit the initial loading data according to a power law relationship:

$$P = \left(h + h_i\right)^m \tag{4.3g}$$

where m and h_i are the unknowns, and P and h are the experimental data. Note that expressed in this way, m in Eq. 4.3g is the inverse of that in Eq. 4.3a. Calculations show that this method and that described by Eq. 4.3e give very similar results.

For a Berkovich indenter, use may be made of the square root relationship between indenter load and penetration depth (Eq. 2.4c) for a completely plastic contact and simply fit a second-order polynomial to the data. Some care should be taken in the event that the bluntness of the tip leads to an initial elastic response, in which case the first method presented here may be used.

4.4 Instrument Compliance

The depth measuring system of a typical indentation instrument registers the depth of penetration of the indenter into the specimen and also any displacements of the instrument arising from reaction forces during loading. These displacements are directly proportional to the load and the general scenario is shown in Fig. 4.2.

The compliance C_f of the loading instrument is defined as the deflection of the instrument divided by the load. The measured unloading stiffness dP/dh during an indentation test has contributions from both the responses of the specimen and the instrument. The contribution from the instrument, C_f, includes the compliance of the loading frame, the indenter shaft, and the specimen mount. The compliance of the indenter material, 1/S, is included in the composite modulus E^* where the stiffness of the contact S is given a rearrangement of Eq. 3.2.3.1i. The specimen/indenter combination and the load frame can be considered as springs in series, in which case, the compliance of each can be added directly to give the total compliance dh/dP measured by the instrument:

$$\frac{dh}{dP} = \frac{1}{S} + C_f \tag{4.4a}$$

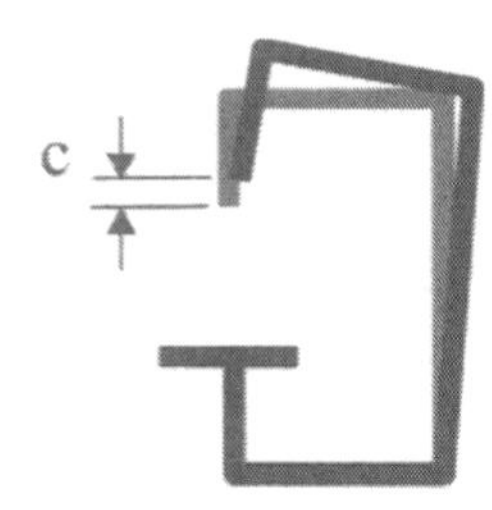

Fig. 4.2 Schematic of the effect of load frame deflection arising from reaction forces during an indentation test. The displacement of the load frame c is measured by the depth measurement system and interpreted as penetration into the specimen material. The magnitude of the deflection is proportional to the load and must be subtracted from the depth readings to obtain the true depth of penetration of the indenter into the specimen.

For the case of a Berkovich indenter, where $A = 24.5h_p^2$, we obtain from Eq. 3.2.3.1i:

$$\frac{dh}{dP} = \sqrt{\frac{\pi}{24.5}}\frac{1}{2\beta E^*}\frac{1}{h_p} + C_f \tag{4.4b}$$

For the case of a spherical indenter, with $A = 2\pi R_i h_p$, we have:

$$\frac{dh}{dP} = \left[\frac{1}{2\beta E^* R_i^{1/2}}\right]\frac{1}{h_p^{1/2}} + C_f \tag{4.4c}$$

The most common method of obtaining a value for C_f, for the case of a spherical indenter, is to plot dh/dP vs $1/h_p^{1/2}$ (or $1/h_p$ for a Berkovich indenter) obtained for an elastic unloading into an elastic–plastic material for a range of maximum indentation depths. This plot should be linear with a slope proportional to $1/E^*$ and an intercept which gives the compliance of the instrument C_f directly. A typical plot is shown in Fig. 4.3.

Experience shows that errors in the data at low values of h_p significantly affect the slope of the fitted line and thus introduce large errors in the estimate of the compliance. It is beneficial to discard a few of the initial data points in the series to obtain the best possible linear fit to the remaining data or to perform a weighted analysis in which less weight is given to data at these lower penetration depths. Since the compliance estimate obtained in this way is dependent upon the area function of the indenter, an iterative procedure is necessary to arrive at a convergent value.

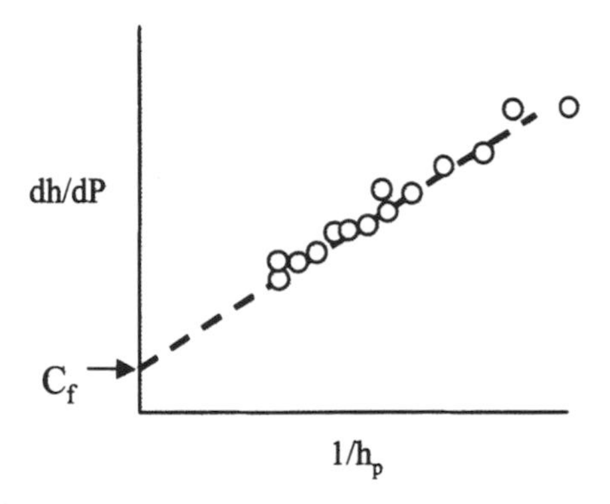

Fig. 4.3 Schematic of dh/dP vs $1/h_p$ (for Berkovich) or $1/h_p^{1/2}$ (for spherical indenter) showing the spread of typical test data and the determination of the compliance C_f from the intercept. Data obtained at low forces, where the uncertainty errors are high, greatly influence the determination of C_f owing to a leverage effect.

In routine data analysis of indentation tests, the compliance of the instrument C_f may be subtracted from experimental values of dh/dP before calculating E^*. Alternately, since the displacements arising from instrument compliance are proportional to the load, a correction may be made to the indentation depths h' (already corrected for initial contact) to give a further corrected depth h" according to:

$$h'' = h' - C_f P \tag{4.4d}$$

An alternative method of establishing instrument compliance involves the testing of a range of specimen materials with a relatively large radius spherical indenter using repeated loading at a single location. The repeated loading minimizes surface effects such as roughness and other irregularities. A relatively large indenter ($R \approx 200$ μm) is used at reasonably high loads where compliance effects are more readily observed and where indenter tip effects (such as non-ideal geometry) are minimized. Since the displacement of the loading column is proportional to the load, the total elastic displacement between two fixed points, remote from the indentation, under a load P is expressed as:

$$\delta = \left[\frac{3}{4E^*}\right]^{2/3} P^{2/3} R^{1/3} + C_f P \tag{4.4e}$$

The displacement δ and load P are measured by the indentation instrument. For any two loads P_1 and P_2 resulting in deflections δ_1 and δ_2, we may form the ratios of Eq. 4.4e and obtain:

$$C_f = \left[\frac{\delta_1 - \left(\frac{P_1}{P_2} \right)^{\frac{2}{3}} \delta_2}{P_1 - \left(\frac{P_1}{P_2} \right)^{\frac{2}{3}} P_2} \right] \tag{4.4f}$$

Note that knowledge of the modulus or hardness of the specimen material is not required in this procedure.

4.5 Indenter Geometry

In nanoindentation testing, the area of contact at penetration depth h_p is found from geometry. The areas A given in Table 1.1 assume that the geometry of the indenter is ideal, a circumstance impossible to achieve in practice. Crystal anisotropy of diamond indenters can also affect the expected shape of the indenter.

Figure 4.4 shows the profile of a nominally 1 μm spherical indenter along with an AFM surface profile of a nominal 10 μm indenter from which the non-ideal character of some indenters are evident. To account for non-ideal geometry of the indenter used in any practical test, it is necessary to apply a correction factor to the equations shown in Table 1.1 so as to determine the real area of contact at a depth h_p. The actual area of contact is given the symbol A and the ideal area of contact for a given value of h_p (that computed from Table 1.1) as A_i. The correction factor to be applied is the ratio A/A_i and is illustrated in Fig. 4.5.

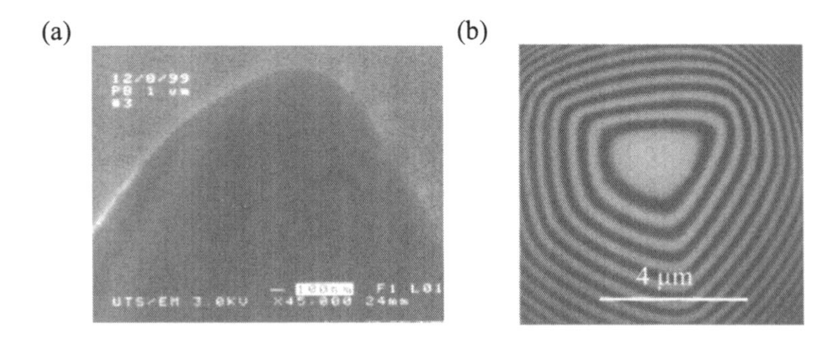

Fig. 4.4 (a) SEM micrograph of the tip of a nominally 1 μm radius diamond sphero-conical indenter used for depth-sensing indentation tests. (b) AFM image showing surface profile of a nominal 10 μm diamond spherical indenter (Courtesy CSIRO).

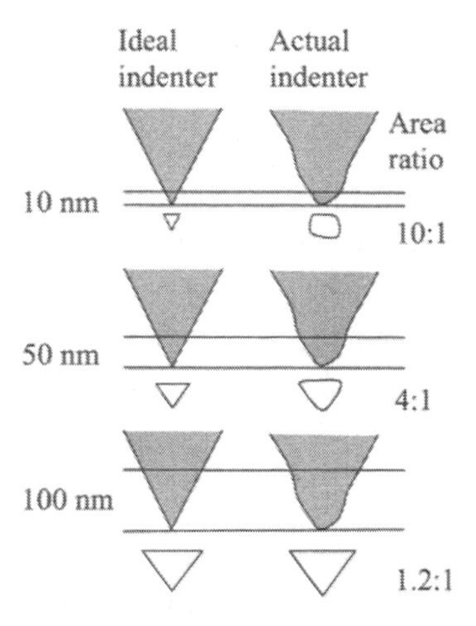

Fig. 4.5 Schematic of comparison of areas of contact with an ideal conical indenter and a real indenter with a non-ideal shape. For the same depth of penetration, the actual area of contact is often larger than that computed with the nominal dimensions of the indenter.

The correction factor can be found from independent measurements of indenter geometry using either an AFM or SEM.[5,6] The measured area A is then plotted against the plastic depth h_p determined from the measured depths (corrected for compliance and initial contact). Regression analysis of the appropriate order may then provide an analytical function that gives the actual projected area for a given value of h_p. This function is commonly called the "area function" for the particular indenter being characterized.

The disadvantage of the direct measurement approach is that is it inconvenient. It is now regular practice to use an indirect method for determining area functions where the procedure is to perform a series of indentations at varying maximum loads on standard test specimens whose elastic modulus and Poisson's ratio are known. If E^* is known (embodying the elastic properties of both indenter and specimen), then the actual area of contact at each load is found from Eq. 3.2.1c thus:

$$A = \pi \left[\frac{dP}{dh} \frac{1}{2\beta E^*} \right]^2 \tag{4.5a}$$

Values of A and h_p for each test on the reference material provide the data for an area function lookup or calibration table. It is often convenient to express the area function as a ratio of the actual area A to the ideal area of indentation A_i. The response for a typical Berkovich indenter is shown in Fig. 4.6. The large value of A/A_i at low values of depth is a consequence of the inevitable bluntness of the indenter tip.

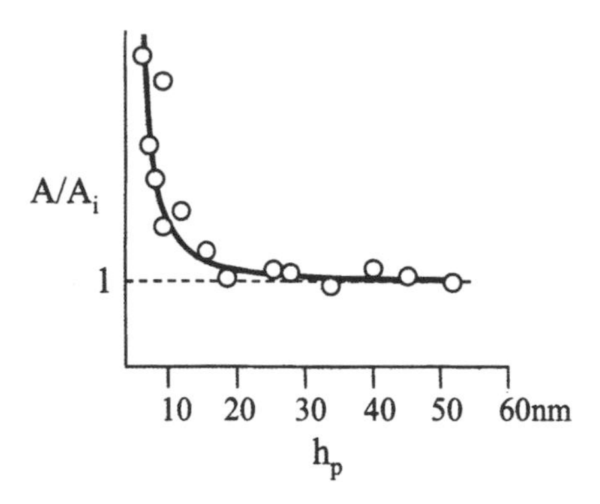

Fig. 4.6 Area correction function for a typical Berkovich indenter. The plot shows the ratio A/A_i as a function of penetration depth h_p. The high value of A/A_i is a consequence of the bluntness of the very tip of the indenter. A/A_i approaches unity as the penetration depth increases.

The corrected hardness is obtained from:

$$H = \frac{P}{A}\left[\frac{A_i}{A}\right] \tag{4.5b}$$

and the corrected modulus is given by:

$$E^* = \frac{dP}{dh}\frac{\sqrt{\pi}}{2\beta\sqrt{A}}\sqrt{\frac{A_i}{A}} \tag{4.5c}$$

where the geometry correction factor β has been included (see Eq. 3.2.4.1f). It should be noted that for a spherical indenter, the contact area is proportional to R. Hence, if the nominal indenter radius is say 5 μm, and the area correction at a particular value of h_p is 0.8, then the effective radius of the indenter at that value of h_p is 4 μm. A value of A/A_i greater than one indicates an indenter with a larger radius than its nominal value. For a given load, this results in a higher value of hardness using uncorrected data in the calculations. A value of A/A_i less than one indicates that the actual area of contact is less than the ideal value, which means that the indenter has a smaller radius than its nominal value. This results in a smaller value of hardness if no correction is applied.

The area correction can be expressed as a mathematical function of the depth h_p in a variety of forms, one popular expression being:

$$A = C_1 h_p^2 + C_2 h_p + C_3 h_p^{1/2} + C_4 h_p^{1/4} + \ldots \tag{4.5d}$$

where the first term represents the ideal area function of the indenter. The ratio A/A_i can thus be calculated where A_i is a function of h_p found from Table 1.1.

In an alternative approach,[7] data from an indentation test on a known specimen can be used to determine the indenter shape function by plotting the square root of the indenter load $P^{1/2}$ (for a conical indenter) or the indenter load P (for a spherical indenter) against the plastic depth h_p. Eqs. 2.3.1b and 2.3.3c show that, for a fully developed plastic zone, the relationship should be linear with the intercept representing the rounding of the tip. Also, by Eq. 3.2.1c, a plot of dP/dh against $P^{1/2}$ should also be linear. These two plots lead to functions of the form:

$$P^n = \frac{h_p - b}{a} \tag{4.5e}$$

$$\frac{dP}{dh} = a\sqrt{P} + b \tag{4.5f}$$

where n = ½ for a conical indenter and n = 1 for a spherical indenter, and where a and b are the slope and intercept terms (different for each plot). Any number of values of the expected values of the indenter load can be calculated for the desired range and spacing of plastic depth terms h_p from Eq. 4.5e and inserted into Eq. 4.5f to give the associated fitted values of dP/dh. Now, from 4.5a, we obtain a value for the actual area of contact and so the ratio A/A_i is:

$$\frac{A}{A_i} = \left(\frac{dP}{dh}\frac{\sqrt{\pi}}{2\beta E^*}\frac{1}{\sqrt{A_i}}\right)^2 \tag{4.5g}$$

where A_i is determined from the selected value of h_p using the formulae given in Table 1.1 for the indenter under consideration and E* is the known combined modulus of the system. This method presumes enough data in the fully plastic regime has been accumulated and as such is best suited for sharp indenters.

4.6 Piling-Up and Sinking-In

In an indentation into an elastic material, the surface of the specimen is typically drawn inwards and downwards underneath the indenter and sinking-in occurs. When the contact involves plastic deformation, the material may either sink in, or pile up around the indenter. In the fully plastic regime, the behavior is seen to be dependent on the ratio E/Y and the strain-hardening properties of the material. The mechanical nature of a typical specimen can be described by a conventional stress–strain relationship that includes a strain-hardening exponent:

$$\begin{aligned} \sigma &= E\varepsilon \qquad \varepsilon \le Y/E \\ \sigma &= K\varepsilon^x \qquad \varepsilon \ge Y/E \end{aligned} \tag{4.6a}$$

where K is equal to:

$$K = Y[E/Y]^x \tag{4.6b}$$

The degree of pile-up or sink-in depends upon the ratio E/Y of the specimen material and the strain-hardening exponent x. Piling-up or sinking-in can be quantified by a pile-up parameter given by the ratio of the plastic depth h_p over the total depth h_t as shown schematically in Fig. 4.7.

For non-strain-hardening materials with a large value of E/Y (e.g., a strain-hardened metal), the plastic zone is observed to have a hemispherical shape meeting the surface well outside the radius of the circle of contact. Piling-up in these materials is to be expected, since most of the plastic deformation occurs in the near the indenter. For materials with a low value of E/Y (e.g., some glasses and ceramics), the plastic zone is typically contained within the boundary of the circle of contact and the elastic deformations that accommodate the volume of the indentation are spread at a greater distance from the indenter. Sinking-in is more likely to occur.

For materials that do exhibit strain-hardening (e.g., a well-annealed metal with $n > 0$), the yield strength effectively increases as its strain increases. Thus, during an indentation test, the material within the plastic zone becomes "harder" as the amount of deformation increases. This means that the outermost material in the plastic zone, which is now "softer," is more susceptible to plastic deformation as the indentation proceeds — the effect is that the plastic zone is driven deeper into the specimen material. Since the material farther away from the indentation is being deformed, the material near the indenter is observed to sink-in as the indenter proceeds downward into the specimen.[8] It is seen therefore that piling-up is most pronounced for non-strain-hardening materials with a high value of E/Y. Sinking-in is more pronounced for strain-hardening or non-strain-hardening materials with a low value of E/Y.

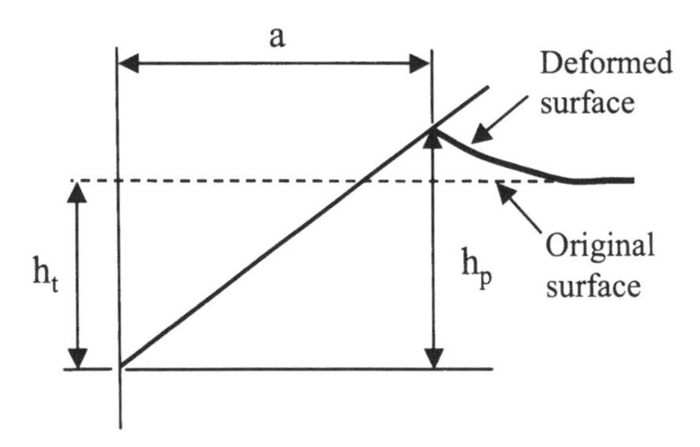

Fig. 4.7 The pile-up parameter is given by h_p/h_t and can be less than or greater than 1.

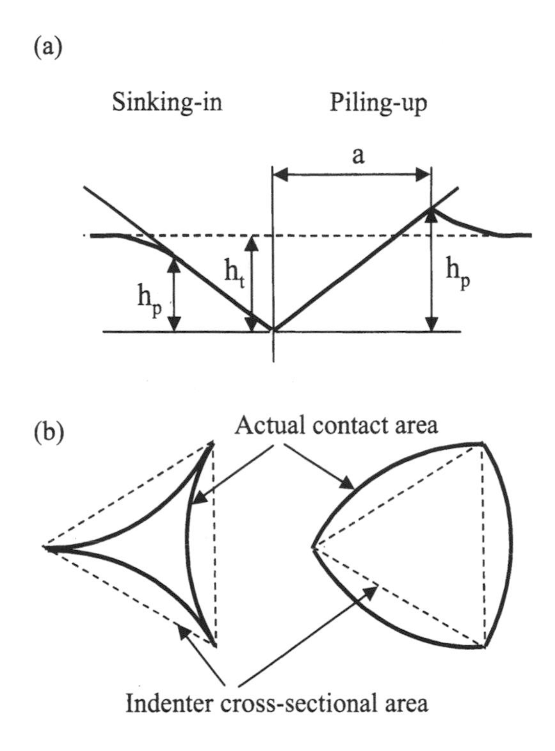

Fig. 4.8 Effect of piling-up and sinking-in on the actual contact area for penetrations of the same depth h_t. (a) Cross-sectional view; (b) plan view. For a given penetration depth h_t, the actual contact area may be substantially different for different materials and to that of the cross-sectional area of the indenter and the expected plastic depth h_p if there were no piling-up or sinking-in. Hay, Oliver, Bolshakov, and Pharr[9] show that there is a unique relationship between the ratio of the slopes of the loading S_L and unloading S_U load-displacement responses with h_r/h_t.

Finite element analysis of contact in which piling-up occurs has demonstrated that the true contact area can be significantly greater than that calculated using the methods of analysis given in previous sections (i.e., from the measured depth of penetration and the assumed elastic unloading response). The effect of piling-up and sinking-in on the contact area is shown schematically in Fig. 4.8. Errors in contact area of up to 60% can be obtained.[8] The existence of piling-up and sinking-in can have a detrimental effect on the determination of the area function of the indenter if the specimen used for determination of the area function behaves differently to that of the eventual specimen to be tested.

Randall and Julia-Schmutz[10] compared the results for mechanical properties obtained from load-displacement curves on gold, titanium, and aluminum-coated silicon wafers with observed features of the residual impression obtained with an AFM. They found that for hard-coated specimens, the indenter load appears to be supported by a combination of elastic flexure and internal stresses of the coating resulting from plastic yielding of the substrate. For soft-coated systems, the indenter cuts through the softer surface layer causing it to be squeezed outwards. The effect on the computed mechanical properties was found to be dependent upon the mode of deformation observed using AFM imaging.

McElhaney, Vlassak, and Nix[11] describe a procedure for accounting for the effects of piling-up and sinking-in based upon measurements of the contact stiffness and SEM pictures of the residual impressions from large indentations. This information provides a correction factor that quantifies the degree of piling-up and or sinking-in and this can be applied to contact depths too small to be readily imaged thus providing a procedure for determination of the area correction function for the indenter.

Bolshakov and Pharr[8] found that the results of finite element analysis showed that ratio of the residual depth h_r to the total depth h_t is a useful parameter for predicting the constraint factor and the extent of pile-up during indentation in bulk materials for a given strain-hardening exponent. The disadvantage of using h_r/h_t for determining the extent of piling-up is that it assumes that the mechanical properties of the specimen are the same at the full penetration depth and the depth of the residual impression. Thus, in bulk materials with uniform mechanical properties, it is found:

$$\frac{S_L}{S_U} = \frac{m_L}{m_U}\left(1 - \frac{h_r}{h_t}\right) \tag{4.6c}$$

where S_L and S_U are the slopes of the loading and unloading curves. The quantity m is a power law exponent that describes the form of the loading and unloading curves. $m_L = 2$ for a geometrically similar indenter (e.g., a cone) and finite element results indicate m_U to be approximately 1.35 for $h_r/h_t > 0.4$. Finite element results[9] indicate the dependence of the true contact area on the ratio S_L/S_U as a result of piling-up or sinking-in if the strain-hardening characteristics of the specimen are known.

4.7 Indentation Size Effect

In a homogeneous, isotropic material, one expects to measure only one value of hardness and modulus, yet, for a variety of reasons, experimental results often result in a variation of hardness and/or modulus with indentation depth. Some of the observed effects are indeed real reflections of material behavior and arise due to the presence of very thin oxide films of substantially different mechanical properties than the bulk material, or the presence of residual stresses and strain-hardening arising from the specimen preparation and polishing procedure. The presence of friction between the indenter and specimen has also been shown to lead to an indentation size effect.[12] The most common observed indentation size effect is probably the errors associated with the area function of the indenter, particularly at very small values of penetration depth. However, even if these effects are minimized, it is still generally observed that for some materials, e.g. crystalline solids, that are nominally isotropic, an indentation size effect is still observed.[13,14]

In materials exhibiting an indentation size effect, the conditions for plastic flow may depend not only on the strain, but also on the magnitude of any strain gradient that might be present in the material. Such gradients exist, for example, in the vicinity of a crack tip, where the stress fields are rapidly changing. Substantial strain gradients also exist in the indentation stress field. In general, the indentation hardness of these materials is observed to increase with decreasing size of indentation owing to the nucleation of dislocations within the plastic zone. Dislocations are created in two ways — those arising for statistical reasons and those arising from the geometry of the indenter. The latter are called geometrically necessary dislocations and take the form of circular dislocation loops as shown in Fig. 4.9.

The presence of dislocations serves to increase the effective yield strength of the material and this in turn means an increase in hardness. Nix and Gao[15] show that the number density of geometrically necessary dislocations created within the plastic zone bounded by the circle of contact for a conical indenter is given by:

$$\rho_g = \frac{3}{2bh}\tan^2\theta \tag{4.7a}$$

In Eq. 4.7a, b is the Burgers vector and θ is the angle of the cone made with the specimen free surface as shown in Fig. 4.9.

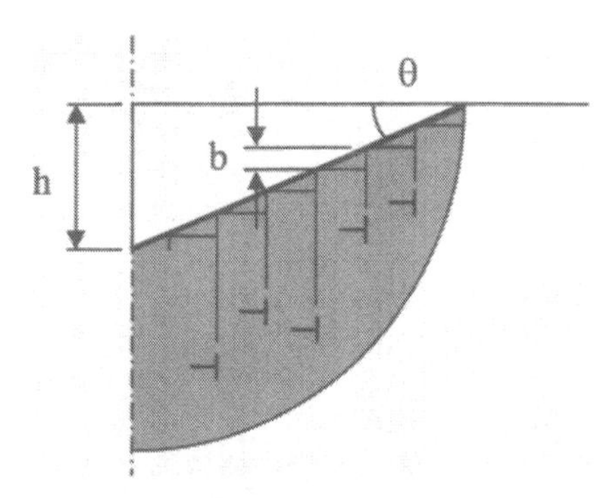

Fig. 4.9 Geometrically necessary dislocations in the plastic zone created by a conical indenter. The plastic zone is contained within the circle of contact.

The significance of Eq. 4.7a is that the density of geometrically necessary dislocations ρ_g increases with decreasing indentation depth h. This leads to an expression for the hardness H in terms of the hardness H_o, which would be obtained without the presence of geometrically necessary dislocations:

$$\frac{H}{H_0} = \sqrt{1 + \frac{h^*}{h}} \qquad (4.7b)$$

where h^* is a length which characterizes the depth dependence of the hardness and which itself depends upon H_o and also ρ_g.

When H^2 is plotted against h, the slope is a measure of h^* and the intercept gives a value for H_o. The significance of Eq. 4.7b is that it shows that the indentation size effect is more pronounced in materials with a low value of intrinsic hardness H_o. For hard materials, very little indentation size effect is expected. For soft material, and especially crystalline materials, a significant indentation size effect is expected. The presence of geometrically necessary dislocations can be explained in terms of the existence of strain gradients in the vicinity of the indentation. The increase in yield strength due to these dislocations becomes more pronounced as the indentation depth becomes smaller, whereupon the strain *gradients* become larger. The existence of an indentation size effect resulting from the presence of dislocations depends on their being sufficient sources of dislocations within the material. Lou, Shrotriya, Buchheit, Yang and Soboyejo[16] used this method for investigating the deformation of LIGA structures where it was found that the microstructural length scale parameter of 2.2 μm was consistent with a gradient scale parameter of 0.9 μm with H_o = 1.64 GPa and h^* = 1.362 μm. Lilleodden, Zimmerman, Foiles and Nix[17] performed an atomic simulation of the nucleation of dislocations in single crystal gold and determined that the existence of high angle grain boundaries serve as a ready source of dislocations in indentation-induced deformation.

Tymiak, Kramer, Bahr, Wyrobek and Gerberich[18] studied single crystals of tungsten and aluminium using a range of conical indenters. They concluded that the magnitude of the strain gradient increased with depth from the specimen surface for spherical indenters and decreased for sharper tipped indenters for both materials yet a factor of two decrease in hardness with increasing depth from the specimen surface was observed in each case.

Zhang and Xu[19] investigated the role of surface effects on nanoindentation hardness at very small penetration depths. They report that there exists a critical indentation depth below which (i.e. shallower) surface effects dominates the load-displacement characteristic. The surface effects consists of energy dissipated at the surface by the indenter load acting against friction stresses and plastic deformation occurring at the surface as well as surface tension. At greater depths, bulk processes predominate.

4.8 Surface Roughness

Surface roughness is a very important issue in nanoindentation. Since the contact area is measured indirectly from the depth of penetration, the natural roughness of real surfaces causes errors in the determination of the area of contact between the indenter and the specimen. There is a large literature on surface roughness since the field is intimately connected with the nature of friction between surfaces.[20–22] The analytical models proposed usually involve elastic or completely plastic contact and often take into consideration the statistical variation in asperity height in real surfaces. Our interest here is the effect of surface roughness on instrumented hardness measurements.

In general, surface roughness is characterized by the asperity height and the spatial distribution of them across the surface. Figure 4.10 shows an AFM scan of an aluminum film deposited onto a glass surface. Average asperity height in this example is in the order of 4 nm.

Surface roughness can be quantified by a roughness parameter α where[23]:

$$\alpha = \frac{\sigma_s R}{a_o^2} \tag{4.8a}$$

In Eq. 4.8a, σ_s is, to a first approximation, equal to the maximum asperity height, R is the indenter radius, and a_o is the contact radius that would be obtained under the same load P for smooth surfaces. Note that α depends indirectly upon the load applied to the indenter. A second parameter μ is also used in combination with α to characterize surface roughness. Johnson[23] finds that the effects of surface roughness on the validity of the elastic contact equations are of significance for $\alpha > 0.05$. The overall effect of surface roughness is to reduce the mean contact pressure by increasing the contact radius. Thus, for a given indenter load P, the depth of penetration is reduced and the computed combined modulus E^* is also reduced.

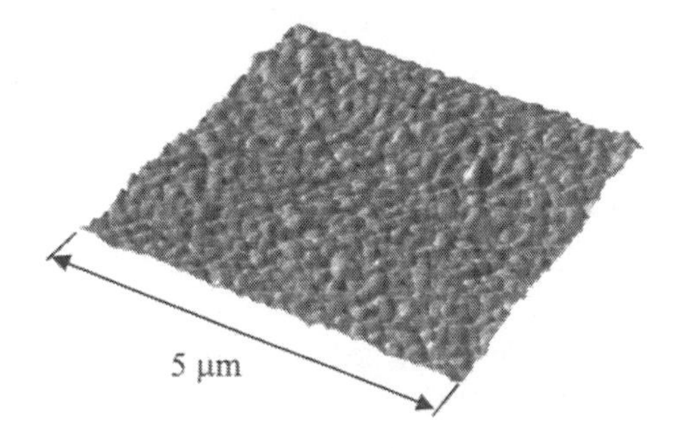

Fig. 4.10 AFM scan of aluminum film on glass. The average "grain size" is estimated to be 300 nm and surface roughness of 4 nm (Courtesy CSIRO).

Equation 4.8a shows that the surface roughness parameter increases with increasing radius of indenter and increases with decreasing indenter load. Thus, for light loads with spherical indenters, surface roughness can be a significant effect. For sharper indenters, say, a Berkovich indenter with a tip radius of 100 nm, the effects of surface roughness are less severe. Joslin and Oliver[24] propose that a new parameter of materials performance, the ratio of H/E^2, is a measure of a material's resistance to plastic deformation, the measurement of which is accomplished by measuring the contact stiffness (dP/dh) and is less sensitive to the effects of surface roughness.

4.9 Tip Rounding

The most commonly used indenter in nanoindentation experiments is the three-sided Berkovich diamond pyramid. As shown in Fig. 4.11, in practice, such indenters are not perfectly sharp but have a tip radius in the order of 100 nm.

Tip rounding becomes important when one wishes to perform indentations on thin films less than about 500 nm thickness and when the maximum depth of penetration is usually of the order of 50 nm. The rounding of the tip results in an initial elastic contact. The resulting hardness measurement may thus be in error since the mean contact pressure may not reflect the condition of a fully formed plastic zone. Modulus measurements are not so much affected by tip rounding as long as the shape of the indenter is well characterized by the area function.

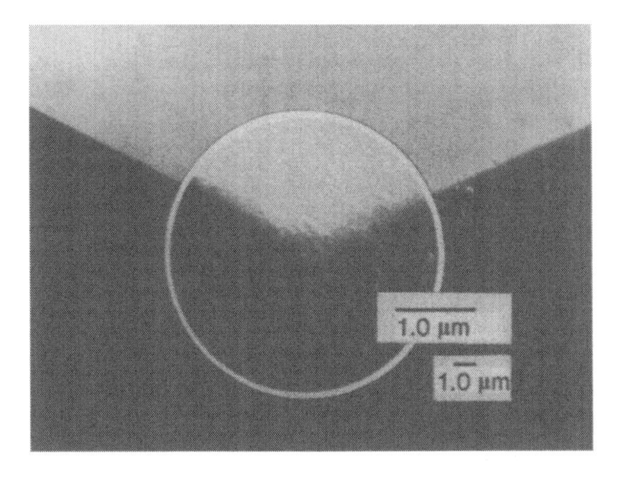

Fig. 4.11 The tip of a Berkovich indenter. Tip radius is on the order of 100 nm. The correspondence in the edges from the center magnified view and the outer region is a fine example of the geometrical similarity associated with this indenter geometry (Courtesy CSIRO and after reference 25).

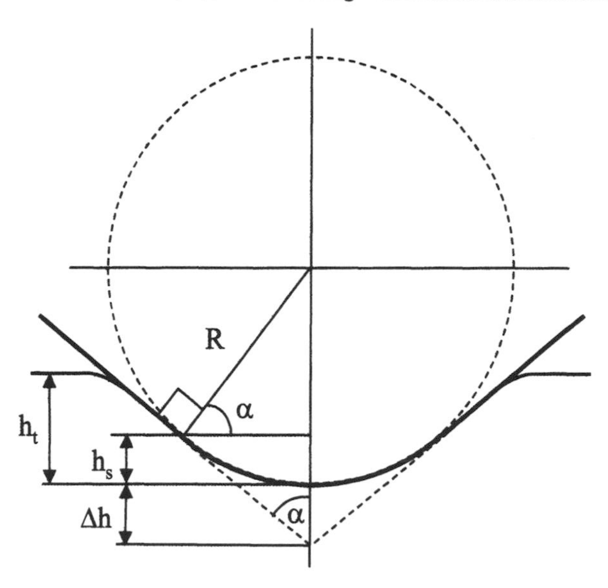

Fig. 4.12 Geometry of a sphero-conical indenter. The conical sides meet the spherical tip at a depth h_s. For a cone angle of $\alpha = 70.3°$, when $h_t/R < 0.73$ the loading is by the spherical tip. For $h_t/R >> 0.73$, the loading is dominated by the cone.

A real indenter is therefore best modeled as a sphero-conical indenter as shown in Fig. 4.12. For a cone of semi-angle α, the depth h_s at which the spherical tip meets the flat face of the cone is given by:

$$h_s = R(1 - \sin\alpha) \tag{4.9a}$$

Thus, the loading with a pyramidal indenter (of equivalent cone angle 70.3°) with a rounded tip of radius R should be identical with that of a spherical indenter for $h_t/R < 0.073$ and only approach that of a sharp indenter when $h_t/R >> 0.073$. In the intermediate regime, indentation with a sphero-cone occurs and so the response is expected to be different to both the cases of spherical and ideally sharp conical indenters. In most cases, the effects of tip rounding are accommodated by the indenter area correction function (see Section 4.5).

If Δh represents the truncation of the tip, then the quantities h_t and h_p are in error by this amount. The significance of this error becomes smaller as the depth of penetration h_t becomes larger. Thus, if it is assumed for the moment that the area function correction accounts for a rounding of the tip, then as h_t gets larger, Δh becomes smaller and the area function A/A_i approaches unity. Since the area function A/A_i for any value of h_p can be measured experimentally, it is possible to calculate the radius of the tip rounding.

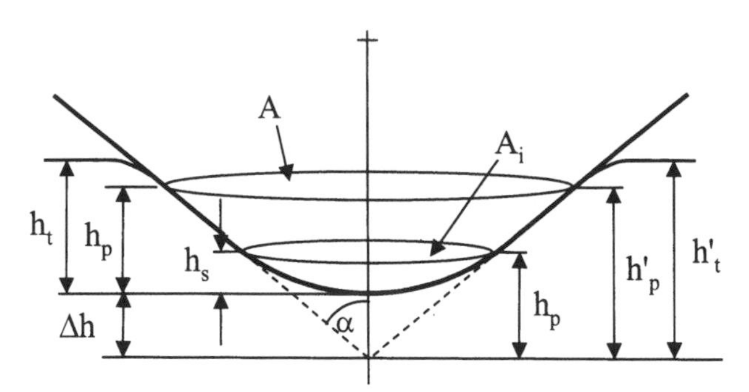

Fig. 4.13 Indentation with a sphero-conical indenter. For a measured value of h_p, the area of contact calculated from geometry is A_i but the real area of contact is a larger value A due to tip rounding. The ratio of the two is the area function A/A_i. Shown above is the special case at which $h_p = h_s$.

As shown in Fig. 4.13, the value h_p is obtained from an experiment, and with no correction applied, the corresponding area of contact would be calculated as:

$$A_{calc} = \pi h_p^2 \tan^2 \alpha \tag{4.9b}$$

The real or actual area of contact is larger than this because of tip rounding and is given by:

$$A_{actual} = \pi h_p'^2 \tan^2 \alpha \tag{4.9c}$$

The ratio of A_{actual} and A_{calc} is the area function A/A_i at that value of h_p.

$$\frac{A_{actual}}{A_{calc}} = \frac{A}{A_i} = \frac{h_p'^2}{h_p^2} \tag{4.9d}$$

Now, consider the special case of $h_p = h_s$, that is, the value of h_p at which the area of contact corresponds to the conjunction of the sphere at the cone.

$$\begin{aligned} h_p' &= h_s + \Delta h \\ h_p &= h_s \end{aligned} \tag{4.9e}$$

Making use of Eqs. 4.9a and 4.9d, it can be shown that at this condition:

$$\frac{A}{A_i} = \left(\frac{1}{\sin \alpha} + 1\right)^2 \tag{4.9f}$$

This value of A/A_i depends only on the cone angle α, and thus is known for any indenter. The corresponding value of h_p at this special condition $h_p = h_s$ can be obtained from the area function A/A_i previously measured using a standard specimen. Now, for an arbitrary load, in which h_p has some general value, the value of h_p' is thus found from:

$$h_p' = \sqrt{h_p^2 \left(\frac{A}{A_i} \right)} \tag{4.9g}$$

The truncation Δh is thus given by:

$$\Delta h = h_p' - h_p \tag{4.9h}$$

Using Eq. 3.2.3.1f and Eq. 4.9a, it can be shown from geometrical considerations that the radius of the tip is given in terms of the truncation Δh and cone angle α by:

$$R = \Delta h \left(\frac{1}{1 - \sin \alpha} - 1 \right) \tag{4.9i}$$

For a typical Berkovich indenter, the tip radius using this method is usually measured to be within the range 75 to 200 nm for new indenters, rising to 200 — 300 nm for well-used indenters.

4.10 Residual Stresses

In the analysis procedures presented in Chapter 3, it was assumed that the specimen material was originally stress-free prior to indentation. In many materials, stresses, tensile or compressive, may be present within the specimen as a result of processing (temperature induced) or surface preparation (cold working from polishing). One way of determining the level of residual stress is to examine the shape of the pile-up occurring at the edge of the contact circle.[26] Deviations in shape give some information about the level and sign of residual stress within the specimen. Other workers have examined the critical load to initiate cone cracks in brittle materials to determine the magnitude and direction of surface residual stresses in brittle materials.[27] Yet another method uses the size of median cracks generated with sharp indenters in brittle solids as a means of determining the level of residual stresses arising from tempering. Chaudhri and Phillips,[28] and Chandrasekar and Chaudhri[29] report that the level of residual stress σ_R can be determined from:

$$\sigma_R = \frac{\chi \left(P^* - P \right)}{1.16 c^2} \tag{4.10a}$$

where P^* and P are the loads that produce cracks of the same radius c in tempered glass with a residual stress and annealed glass, respectively, and, for a Vickers indenter, where:

$$\chi = \frac{1}{\pi^{3/2} \tan 68} \tag{4.10b}$$

Deviations in shape of the load-displacement response from the ideal shape could also be used an indication of residual stresses but experiments show that the effect is too small to be measured accurately[30] although other studies[31] generally report a smaller value of depth h_t for a maximum load P_t for a compressive stress and a larger value for a tensile stress compared to the stress-free state. Lee and Kwon[32] demonstrate that the residual stress for a thin film can be found from a quantitative difference between the load-displacement curves for stressed and stress-free specimens. For a given depth of penetration, the load for the stress-free state may be given the symbol L_o. For a tensile stress, the load at this depth is a lower load L_T and for a compressive stress, the load is L_C. The difference is L_{Res} ($= L_o - L_T$ for tension, $L_o - L_C$ for compression) and the magnitude of the residual stress is found from:

$$\sigma_{Res} = \frac{L_{Res}}{A} \tag{4.10c}$$

where A is the area of contact. Comparison with a wafer curvature method showed good correlation although the stress calculated from Eq. 4.10c was observed to decrease with decreasing indenter load.

More recently, Taljat and Pharr,[33] who compared the features of experimentally derived load-displacement responses obtained with a spherical indenter to those obtained by finite element calculation within which residual stresses could be incorporated to various degrees in a controlled manner. In this method, it is recognized that for an elastic contact, any residual stress in the specimen serves to alter the stress distribution from the Hertzian case and thus causes a change in the indentation strain (a/R) at which yield might first occur. In Chapter 2, it was found that yield first occurs in the Hertzian stress field when the mean contact pressure is approximately equal to 1.1 times the material yield stress. If we then add a residual biaxial stress, we obtain, at the onset of yielding:

$$p_m = 1.1(Y - \sigma_R) \tag{4.10d}$$

which, when combined with the Hertz equation, Eq. 1.2j, yields:

$$\sigma_R = Y - \left(\frac{4E^*}{3.3\pi}\right)\frac{a}{R} \tag{4.10e}$$

Equation 4.10e allows σ_R to be determined if Y and E^* are known along with a/R at the onset of yielding. Comparison with finite element results over a wide range of E, Y, and residual stresses show that the elastic recovery parameter h_r/h_t

has a unique correlation with residual stress for a particular value of the quantity $E/Y\ 2h_t/a_t$, where a_t is the radius of the indenter measured in line with the original specimen free surface.

In a further development of this method, Swadener, Taljat and Pharr[34] noticed that, for indentations with a spherical indenter, when the mean contact pressure p_m is plotted against the parameter Ea/YR within the transition region, the data obtained was offset vertically by an amount equivalent to the magnitude of the applied biaxial stress leading to:

$$p_m + \sigma_R = C\sigma_F \tag{4.10f}$$

where σ_F is the flow stress given by the difference between the yield stress and the residual stress ($\sigma_F = Y - \sigma_R$). σ_R can thus be determined if experiments are performed at different values of p_m within the transition region where the value of the constraint factor C is changing. The procedure requires calibration against a prepared specimen in a known state of stress to determine the manner in which C varies with p_m in the transition region for the type of material being studied. It does however have the advantage of not requiring a value of yield stress Y to be known in advance.

4.11 Specimen Preparation

It is evident from Section 4.8 that surface roughness is an important issue for nanoindentation tests. In order to improve the quality or the specimen surface, it is common practice to polish the specimen. Specimen polishing is a specialized activity and requires considerable care and experience for an acceptable result.

Polishing is usually done by holding the specimen in contact with a rotating polishing wheel upon which is placed a mat that has been impregnated with a polishing compound — usually a suspension of fine particles in a lubricant. A progressive decrease in grit size, with thorough rinsing in-between to minimize contamination, is usually required. A 1 μm polishing compound usually results in what appears to be a very good mirror finish.

One important by-product of the polishing procedure, especially for metals, is the modification of the surface of the specimen due to strain-hardening or cold-working. The polishing procedure involves a substantial amount of deformation of the surface of the specimen material, and it is common to encounter an unwanted indentation size effect resulting from the polishing procedure. Langitan and Lawn[35] found that the resulting flaw size from polishing brittle materials is approximately half the nominal grit size. Allowing for the extended size of the plastic zone in ductile materials, it is reasonable to assume therefore that the polishing procedure affects the surface of the specimen to a depth of about the same size as the nominal grit size. For metal specimens that are likely to strain-harden, one could therefore expect to find a change in mechanical properties in the specimen over this depth range.

References

1. N.M. Jennett and J. Meneve, "Depth sensing indentation of thin hard films: a study of modulus measurement sensitivity to indentation parameters," Mat. Res. Soc. Symp. Proc. 522, 1998, pp. 239–244.
2. G. Feng and A.H.W. Ngan, "Effects of creep and thermal drift on modulus measurement using depth-sensing indentation," J. Mater. Res. 17 3, 2002, pp. 660–668.
3. A.C. Fischer-Cripps, unpublished work, 2003.
4. B.R. Lawn, B.J. Hockey and S.M. Weiderhorn, "Thermal effects in sharp-particle impact," J. Amer. Ceram. Soc. 63 5–6, 1980, pp. 356–358.
5. K. Herrmann K, N.M. Jennett, W. Wegener, J. Meneve, K. Hasche, and R. Seemann, "Progress in determination of the area function of indenters used for nanoindentation," Thin Solid Films, 377, 2000, pp. 394–400.
6. S. Enders, P. Grau, and H.M. Hawthorne, "Determination of the real indenter shape for nanoindentation/nanotribology tests by surface metrological and analytical investigations," Mat. Res. Soc. Symp. Proc. 649, 2001, pp. Q3.6.1–Q3.6.6.
7. R.N. Bolster, unpublished work, 2001.
8. A. Bolshakov and G.M. Pharr, "Influences of pileup on the measurement of mechanical properties by load and depth sensing indentation techniques," J. Mater. Res. 13 4, 1998, pp. 1049–1058.
9. J.L. Hay, W.C. Oliver, A. Bolshakov, and G.M. Pharr "Using the ratio of loading slope and elastic stiffness to predict pile-up and constraint factor during indentation," Mat. Res. Proc. Symp. 522, 1998, pp. 101–106.
10. N. X. Randall and C. Julia-Schmutz, "Evolution of contact area and pile-up during the nanoindentation of soft coatings on hard substrates." Mat. Res. Soc. Symp. Proc. 522, 1998, pp. 21–26.
11. K.W. McElhaney, J.J. Vlassak, and W.D. Nix, "Determination of indenter tip geometry and indentation contact area for depth-sensing indentation experiments," J. Mater. Res. 13 5, 1998, pp. 1300–1306.
12. H. Li, A. Ghosh, Y.H. Yan, and R.C. Bradt, "The frictional component of the indentation size effect in low load microhardness testing," J. Mater. Res. 8 5, 1993, pp. 1028–1032.
13. N. Gane, "The direct measurement of the strength of metals on a sub-micrometre scale," Proc. Roy. Soc. A317, 1970, pp. 367–391.
14. S.J. Bull, T.F. Page, and E.H. Yoffe, "An explanation of the indentation size effects in ceramics," Phil. Mag. Lett. 59 6, 1989, pp. 281–288.
15. W.D. Nix and H. Gao, "Indentation size effects in crystalline materials: a law for strain gradient plasticity," J. Mech. Phys. Solids, 46 3, 1998, pp. 411–425.
16. J. Lou, P. Shrotriya, T. Buchheit, D. Yang and W.O. Sobojeyo, "Nanoindentation study of plasticity length scale effects in LIGA Ni microelectromechanical systems structures," J. Mater. Res. 18 3, 2003, pp. 719–728.
17. E.T. Lilleodden, J.A. Zimmerman, S.M. Foiles, and W.D. Nix, "Atomistic simulations of elastic deformation and dislocation nucleation during nanoindentation," J. Mech. Phys. Solids, 51, 2003, pp. 201–920.

18. N.I. Tymiak, D.E. Kramer, D.F. Bahr, T.J. Wyrobek and W.W. Gerberich, "Plastic strain and strain gradients at very small indentation depths," Acta Mater. 49, 2001, pp. 1021–1034.
19. T.-Y. Zhang and W.-H. Zu, "Surface effects on nanoindentation," J. Mater. Res. 17 7, 2002, pp. 1715–1720.
20. J.F. Archard, "Elastic deformations and the law of friction," Proc. Roy. Soc. A243, 1957, pp. 190–205.
21. J.A. Greenwood and J.B.P. Williamson, "Contact of nominally flat surfaces," Proc. Roy. Soc. A295, 1966, pp. 300–319.
22. J.A. Greenwood and J.H. Tripp, "The contact of two nominally rough surfaces," Proc. Inst. Mech. Eng. 185, 1971, pp. 625–633.
23. K.L. Johnson, *Contact Mechanics*, Cambridge University Press, Cambridge, 1985.
24. D.L. Joslin and W.C. Oliver, "A new method for analyzing data from continuous depth-sensing microindentation tests," J. Mater. Res. 5 1, 1990, pp. 123–126.
25. J.S. Field, "Understanding the penetration resistance of modified surface layers," Surf. Coat. Tech. 36, 1988, pp. 817–827.
26. J.H. Underwood, "Residual stress measurement using surface displacements around an indentation," Experimental Mechanics, 30, 1973, pp. 373–380.
27. S.G. Roberts, C.W. Lawrence, Y. Bisrat, and P.D. Warren, "Determination of surface residual stresses in brittle materials by Hertzian indentation: Theory and experiment," J. Am. Ceram. Soc. 82 7, 1999, pp. 1809–1816.
28. M.M. Chaudhri and M.A. Phillips, "Quasi-static cracking of thermally tempered soda-lime glass with spherical and Vickers indenters," Phil. Mag. A 62 1, 1990, pp. 1–27.
29. S. Chandrasekar and M.M. Chaudhri, "Indentation cracking in soda-lime glass and Ni-Zn ferrite under Knoop and conical indenters and residual stress measurements," Phil. Mag. A 67 6, 1993, pp. 1187–1218.
30. T.Y. Tsui, W.C. Oliver, and G.M. Pharr, "Influences of stress on the measurement of mechanical properties using nanoindentation. 1. Experimental studies in an aluminium alloy," J. Mater. Res. 11 3, 1996, pp. 752–759.
31. A. Bolshakov, W.C. Oliver, and G.M. Pharr, "Influences of stress on the measurement of mechanical properties using nanoindentation. 2. Finite element simulations," J. Mater. Res. 11 3, 1996, pp. 760–768.
32. Y.-H. Lee and D. Kwong, "Residual stresses in DLC/Si and Au/Si systems: Application of a stress-relaxation model to nanoindentation technique," J. Mater. Res. 17 4, 2002, pp. 901–906.
33. A. Taljat and G.M. Pharr, "Measurement of residual stresses by load and depth sensing spherical indentation," Mat. Res. Soc. Symp. Proc. 594, 2000, pp. 519–524.
34. J.G. Swadener, B. Taljat, and G.M. Pharr, "Measurement of residual stress by load and depth sensing indentation with spherical indenters," J. Mater. Res. 16 7, 2001, pp. 2091–2102.
35. F.B. Langitan and B.R. Lawn, "Hertzian fracture experiments on abraded glass surfaces as definitive evidence for an energy balance explanation of Auerbach's law," J. App. Phys. 40 10, 1969, pp. 4009–4017.

Chapter 5
Simulation of Nanoindentation Test Data

5.1 Introduction

The methods of analysis described in Chapter 3 can be used to provide a useful computation of simulated load-displacement curves, where the mechanical properties of both the specimen and indenter are given as input parameters. A simulated load-displacement curve allows comparisons to be made with actual experimental data. For example, such comparisons may yield information about non-linear events such as cracking or phase changes that might occur with an actual specimen during an indentation test. In this chapter, the procedure for generating a simulated load-displacement curve is described in detail and a comparison is made with experimental data from materials with a wide range of ratio of modulus to hardness.

5.2 Spherical Indenter

Upon the application of load during an indentation test with a spherical indenter, there is generally an initial elastic response followed by elastic plastic deformation. Following Field and Swain,[1] for the purposes of simulation, the transition between the two responses is assumed to occur at a mean contact pressure equal to the hardness H. That is, it is assumed that there is an abrupt transition from elastic deformation to a fully developed plastic zone and no intermediate region. The data to be calculated are the load and depth for both loading and unloading sequence. At low loads, the specimen response is elastic and the relationship between depth and load is given by Eqs. 3.2.3a and, in terms of the contact radius, 3.2.3.2a. When a critical load P_c is reached, full plasticity is assumed and $p_m = H$. Thus, with $p_m = P/\pi a^2$, it can be shown from Eq. 3.2.3.2a that the critical load is given by:

$$P_c = \left(\frac{3}{4E^*}\right)^2 (\pi H)^3 R_i^2 \qquad (5.2a)$$

At the critical load P_c, there is a corresponding radius of circle of contact a_c which can be expressed in terms of the hardness H:

$$a_c = \sqrt{\frac{P_c}{\pi H}} \tag{5.2b}$$

Beyond the critical load, it is assumed that no further increase in the mean contact pressure occurs with increasing depth (H is a constant). The contact is now considered "plastic." In this plastic region, it is desired to calculate the depth of penetration beneath the specimen free surface h as a function of indenter load P given values of E^* and H for the material. The first step is to calculate the radius of the circle of contact from the hardness and the load in terms of the radius of the circle of contact at the critical load:

$$a = a_c\left(\frac{P_t}{P_c}\right)^{\frac{1}{2+x}} \tag{5.2c}$$

where x is the strain-hardening exponent. For a constant value of hardness, or a perfectly elastic–plastic material, x = 0. Upon unloading, it can be assumed that the response is elastic from the depth h_t at full load P_t, to the final residual depth h_r and the elastic displacement is h_e given by Eq. 3.2.3.2a. From Eq. 3.2.3.2b, we obtain:

$$h_t = h_p + \frac{h_e}{2} \tag{5.2d}$$

and thus:

$$h_t = \frac{1}{2}\left(2h_p + \frac{3}{4E^*}\frac{P_t}{a}\right) \tag{5.2e}$$

where h_p and a are found from the hardness value H via Eqs. 3.2.3d and 3.2.3.1f to give:

$$h = \frac{1}{2}\left(\frac{P}{\pi R_i H} + \frac{3}{4}\frac{\sqrt{P\pi H}}{E^*}\right) \tag{5.2f}$$

Equation 3.2.3.2a provides the loading curve up to a critical load P_c after which full plasticity is assumed and Eq. 5.2f is then appropriate.

The unloading curve is assumed to describe a fully elastic response from the total depth h_t to a residual depth h_r and, hence, the elastic displacement given by Eq. 3.2.3a is to be added to the residual depth h_r to given the total value of h_t upon unloading. However, the unloading involves that between the indenter and the residual impression of radius R_r and hence R in Eq. 3.2.3a is the combined radius of curvature R given by Eq. 3.2.3b. Now, R_r is unknown, but since the radius of the circle of contact a at full load and the depth h_r are known R_r can be determined from Eq. 3.2.3.1f thus:

$$R_r \approx \frac{a^2}{2h_r} \tag{5.2g}$$

The penetration depth beneath the surface of the residual impression is thus given by Eq. 3.2.3a with R, the relative curvatures, given by Eq. 3.2.3b. The absolute value of the unloading penetration (from the original specimen free surface) is this value of h_e added to the depth of the residual impression h_r.

5.3 Berkovich Indenter

For the case of a Berkovich indenter, a completely elastic–plastic response can be assumed. The loading curve can be found from the addition of h_p and h_a as shown in Fig. 3.4 where, from Eq. 1.2m and Table 1, we have:

$$h_p = \left(\frac{P}{3\sqrt{3}H\tan^2\theta}\right)^{\frac{1}{2}} \tag{5.3a}$$

The distance h_a is most easily determined from the intercept of the slope of the unloading curve at maximum load P_t as shown in Fig. 3.4.

$$h_a = \left[\frac{2(\pi-2)}{\pi}\right]\frac{1}{2E^*}(P_t H\pi)^{1/2} \tag{5.3b}$$

The total depth h_t is thus:

$$h_t = \sqrt{P_t}\left[\left(3\sqrt{3}H\tan^2\theta\right)^{-\frac{1}{2}} + \left[\frac{2(\pi-2)}{\pi}\right]\frac{\sqrt{H\pi}}{2E^*}\right] \tag{5.3c}$$

Equation 5.3c is applicable at any load P giving a depth of penetration h and so the t subscripts can be discarded if desired.

On unloading, the response is elastic from h_t to the depth of the partial unload h_s. The depth associated with P_t can be calculated from Eq. 5.3c. All that is required is the corresponding depth h_s, where we are free to choose a value of P_s. Rearranging Eq. 3.2.4.2d, we obtain:

$$h_s = \left(\frac{P_s}{P_t}\right)^{1/2}\left[h_r\left(\left(\frac{P_t}{P_s}\right)^{1/2} - 1\right) + h_t\right] \tag{5.3d}$$

where h_r is found from a rearrangement of Eq. 3.2.4.2c.

It should be noted that for a general conical indenter of half-angle α, Eq. 5.3c can be written:

$$P = E^*\left[\frac{1}{\sqrt{\pi}\tan\alpha}\sqrt{\frac{E^*}{H}}+\left[\frac{2(\pi-2)}{\pi}\right]\sqrt{\frac{\pi}{4}}\sqrt{\frac{H}{E^*}}\right]^{-2} h^2 \qquad (5.3e)$$

which illustrates the $P \propto h^2$ relationship for elastic–plastic loading more clearly. Equation 1.2m shows a similar $P \propto h^2$ relationship applicable to the elastic unloading portion of the test cycle.

5.4 Other Indenters

The analyses above may be readily applied to indenters of other geometries. Table 1.1 shows the relevant expressions for a range of common indenter geometries. Note that for the case of the Berkovich, Knoop, and cube corner indenters the intercept correction is given as 0.75 rather than 0.72 as found by Oliver and Pharr.[2] For axial-symmetric indenters, the geometry correction factor is 1.0 while for the others, it is that given by King.[3] It is of interest to note that the analyses given above for what are nominally conical indenters (i.e., Berkovich, Knoop, Vickers, cube corner, and cone) assume a condition of full plasticity from the moment of contact with the specimen. For indenters of ideal geometry, this is perfectly reasonable as the stress singularity at the indenter tip would ensure plastic deformation from the moment of contact. In practice, there is inevitable blunting of the indenter tip that would lead to some small initial elastic response.

There is an interesting issue with the use of the elastic equations that requires consideration. For elastic contact with a conical indenter, the mean contact pressure is given by:

$$p_m = \frac{E^*}{2}\cot\alpha \qquad (5.4a)$$

which is independent of load. The significance of this is that, in the simulation analyses, it is assumed that the mean contact pressure is limited to the hardness value H. However, if the combination of E and the angle α are such that the mean contact pressure given by Eq. 5.4a falls below the specified hardness value H, then the contact is entirely elastic. The elastic equations provide a value of mean contact pressure, even with the stress singularity at the indenter tip. The assumption in the simulation modeling is that a fully developed plastic zone occurs when the mean contact pressure becomes equal to the hardness H, and in the case of the sphere, this happens at a critical load P_c as given by Eq. 5.2a and the mean contact pressure increases with increasing load. In the case of a cone, the mean contact pressure is independent of load and, if it is less than the specified hardness value H, the load-displacement response must be assumed to be entirely elastic with a zero residual depth for all values of load[4].

5.5 Finite Element Analysis

With the ready availability of finite element analysis programs it is appropriate to consider a numerical approach to the simulation of load displacement responses for materials undergoing nanoindentation.

The elastic stress fields generated by an indenter, whether it be a sphere, cylinder, or cone, although complex, are well defined. When the response of the specimen material is elastic–plastic, however, theoretical treatments are limited because of the simplifying assumptions required to make such analyses tractable. IN this case, finite element analysis provides valuable information about the state of stress within the specimen material during both loading and unloading.

For frictionless contact, the contact area is completely specified in terms of linear elasticity as embodied in the equations presented in Chapter 3. Thus, as far as finite-element modeling is concerned, the same solution should be obtained whether the load is applied in one step or as a series of increments. The foregoing discussion assumes that all displacements are small. If this is not the case, then the elastic equations do not apply, and for finite-element modeling, nonlinear *geometric* considerations must be included in the analysis.

The solution of contact problems by finite-element analysis is often conveniently undertaken with the use of specialized gap elements. Although for frictionless contact, a linear solution is obtained if the full indenter load is applied at once, modeling of the expanding area of contact, such as that required for spherical and conical indenters, does require an iterative procedure to account for the development of the plastic zone. In the iterative procedure, the finite element program continually checks the status of each gap element, deleting and reinstating the element as required, until force equilibrium is reached within a specified tolerance level. Ideally, the gap elements should be assigned an infinite stiffness. This would ensure a non-intrusive contact between the indenter and the specimen. However, this is not possible in practice due to the finite numeric restrictions imposed by computer hardware. During contact, the indenter intrudes the specimen a small amount, given by the penalty factor. The penalty factor is the ratio of the stiffness of the gap elements to the stiffness of the specimen material. It is desirable at least to have the stiffness of the gap beams considerably larger than that of the specimen material. If the penalty factor is too low, then there is insufficient stiffness to enforce the contact condition. A penalty factor of $\approx$10000 is usually sufficient.

For material non-linearity, the local elastic modulus of each element is modified at each iteration for each load increment to satisfy a specified constitutive relationship. For example, the shear-driven nature of the subsurface contact damage observed in metals and some structural ceramics would suggest that the

Tresca or von Mises shear stress yield criterion may be specified in the finite-element procedure. The elastic–plastic properties of the specimen material in the finite-element model are specified by a uniaxial stress-strain relationship. In the absence of detailed information about a particular specimen, an elastic, perfectly plastic material response usually gives adequate results.

To account for the localized deformation of the specimen material, it is essential that the density of nodes at and near the contact be high in comparison to the depth of penetration being modeled. Further, the outer boundaries of the model should be placed far enough away so as to represent an infinite half space unless a particular geometry is being modeled (e.g. a thin film). Fig. 5.1 shows a typical finite element mesh where element grading has been implemented to increase the density of nodes at the contact region.

Considerable care needs to be taken in the interpretation of the results of a finite-element analysis. One of the issues that requires caution is the orientation of the elements comprising the model. Elements in a finite-element model generally have a local coordinate system, which may not coincide with the overall, or global, coordinate system of the geometry of the structure. Local coordinate systems are often used to specify anisotropic material properties. The nature of the geometry of the structure may require that the stresses obtained from a finite-element solution be expressed relative to planes aligned with each element's local coordinate axes or with respect to the global coordinate axes.

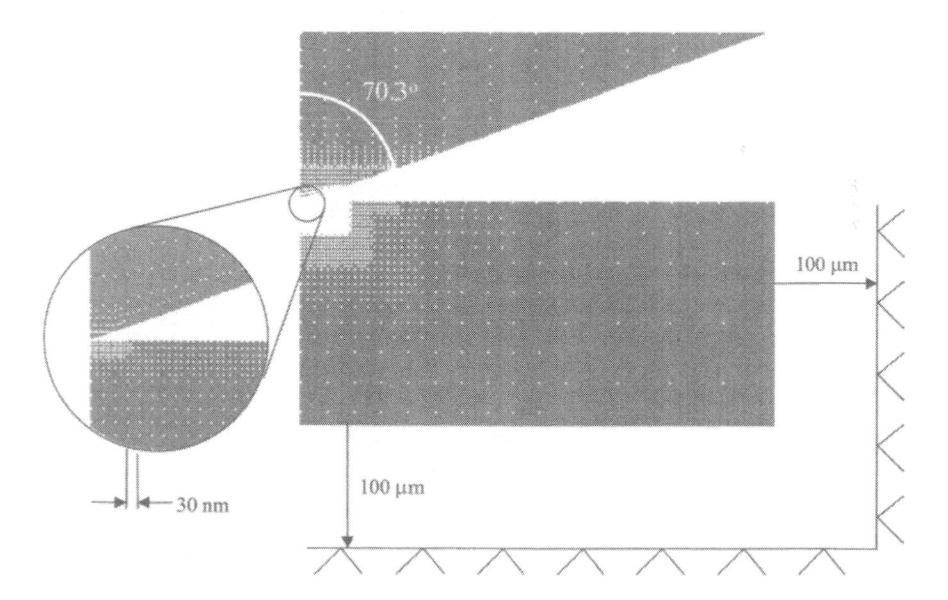

Fig. 5.1 Finite element mesh used for elastic plastic loading and unloading of a half-space (Courtesy CSIRO and after reference 5).

Another potential problem is the averaging technique employed by the particular finite-element analysis code. Typically, stresses, strains, and displacements are reported for each node. However, internally, the program may actually calculate the solutions at the centroids of each area shown and provide a weighted average nodal output. If the nodes of interest lie on a boundary (say between elements with two different material properties), then this averaging may reduce the actual stresses at the boundary where the results are required. In some cases, it may be possible to specify that averaging is performed only within elements of the same material property.

Often, a simple display of stresses is sufficient for a particular analysis. Stress concentrations may be readily identified and future modeling performed on a revised geometry or loading as required. For nanoindentation applications, the load-displacement response is of particular importance and this can readily be obtained directly from nodal displacements in the finite element solution.

5.6 Comparison of Simulated and Experimental Data

It is of interest to evaluate the applicability of the simulation equations by comparison with experimental data. Figures 5.2 and 5.3 show the results from experiments on fused silica and hardened steel together with the load-displacement curve predicted by the simulation methods described above and also elastic–plastic finite element analysis of the contact for the case of a Berkovich indenter[5]. The values of modulus and hardness calculated from the experimental data, along with the maximum load in each experiment, were used as inputs for the simulation calculation and finite element analysis. For the finite element analysis, an axis-symmetric cone of semi-angle 70.3° was modeled, this angle giving the same area to depth relationship to a 3-sided pyramidal Berkovich indenter.

For the loading portion of the load-displacement curves, there is good agreement between the data obtained experimentally, and that computed using the finite element method and also that computed from Eq. 5.3e. Even though the unloading curves for the finite element result and the experiment are nearly parallel, there is an appreciable offset but not too much significance should be attached to this since the finite element (and the theoretical results) depend on the material acting in a perfectly elastic–perfectly plastic manner (with no strain-hardening) that is not strictly representative of that exhibited by the material in practice, especially for steel, where a gradual transition from elastic to plastic response is more to be expected. It is interesting to note that the agreement between all three curves for the fused silica specimen is quite good – a specimen in which the transition from elastic to plastic response is likely to be more marked and with no expected strain-hardening. The good agreement between the theoretical and finite element results for both materials indicate that the choice of constraint factor above was appropriate for each material, the finite element results being dependent on the choice of Y, while the theoretical results depend upon the choice of H.

It is now appropriate to consider the unloading half of the load-displacement response. The unloading response in depth-sensing indentation testing is commonly held to be completely elastic and this assumption underlies the conventional methods of determining hardness and modulus. An examination of the Tresca stress fields in the finite element models used here show, that for this combination of modulus and yield stress, no plastic deformation (reverse plasticity) occurs during unloading. Reloading the residual impression is thus expected to retrace the unloading load displacement response. As shown in Figs. 5.2 and 5.3, the experimental for the reloading of the residual impression demonstrate that this is the case. Finite element results (note shown in Figs. 5.2 and 5.3) for the reloading of the residual impression confirm this observation.

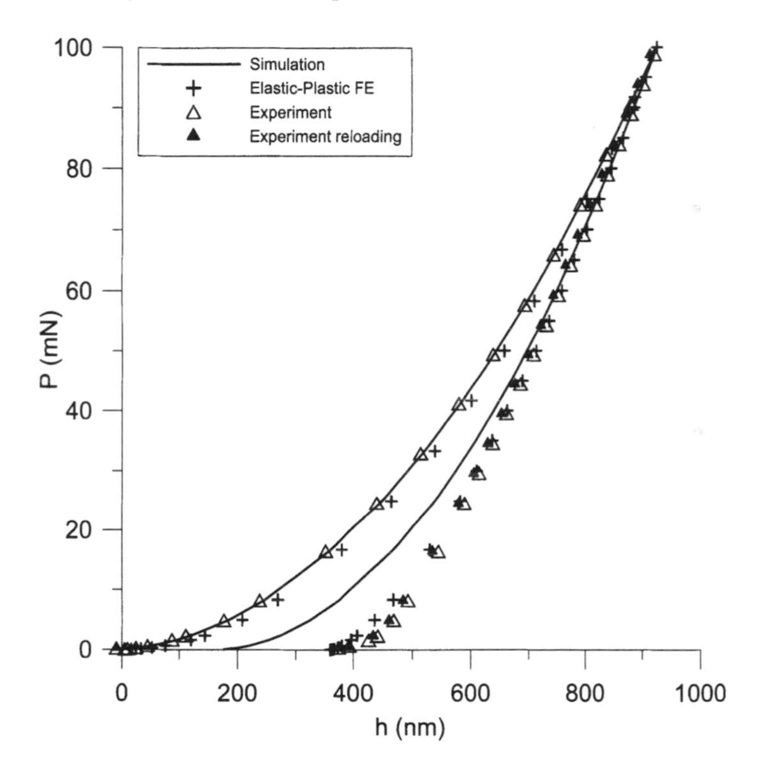

Fig. 5.2 Load-displacement curve for Berkovich indenter on fused silica. The solid line is the results of the theoretical simulation for E = 72.5 GPa and H = 9.53 GPa. Open triangles indicate the experimental load-displacement curve, closed triangles indicate the experimental reloading of the residual impression. Crosses indicate the results of elastic–plastic finite element analysis assuming a rigid indenter and specimen modulus (E^*) set to 69.7 GPa with a yield stress Y = 6.6 GPa and a constraint factor of 1.5 (Courtesy CSIRO and after reference 5).

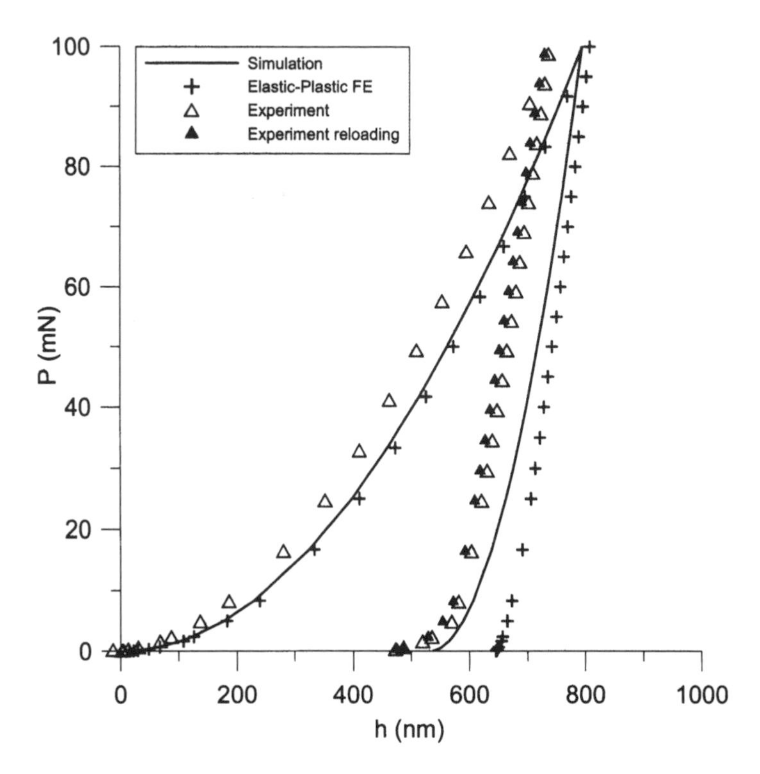

Fig. 5.3 Load-displacement curves for hardened steel with a Berkovich indenter. The solid line is the results of the theoretical simulation for E = 210 GPa and H = 9.6 GPa. Open triangles indicate the experimental load-displacement curve, closed triangles indicate the experimental reloading of the residual impression. Crosses indicate the results of elastic–plastic finite element analysis assuming a rigid indenter and specimen modulus (E^*) set to 187 GPa with a yield stress Y = 3 GPa and a constraint factor of 3 (Courtesy CSIRO and after reference 5).

Despite the observations above, Figs. 5.2 and 5.3 clearly show that there is some substantial difference in the shape of the unloading curves as obtained by experiment, using the finite element method and the theoretical response. The reason for this discrepancy is clearly evident from the shape of the deformed surface of the finite element model. Fig. 5.4 shows that the profile of the residual impression is curved, and not that of a right circular cone as assumed by the theoretical analysis. Similar observations were made on PMMA (a highly elastic solid) by Stilwell and Tabor[6].

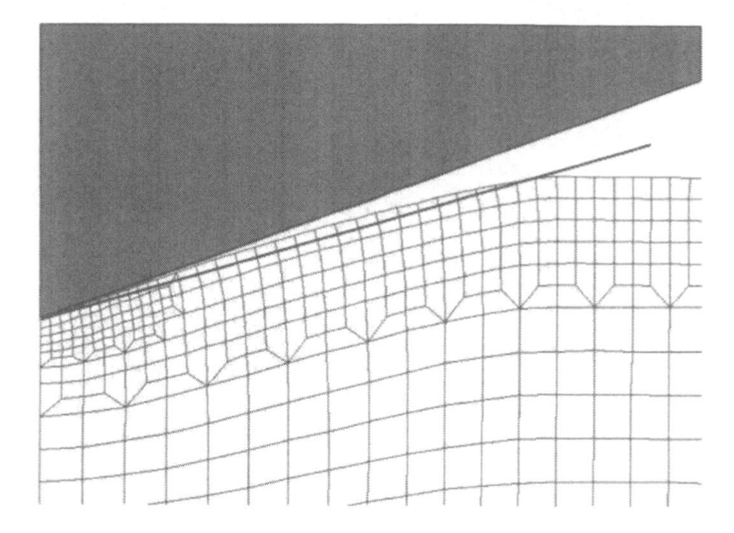

Fig. 5.4 Section view of finite element results of the shape of the residual impression in specimen surface for fused silica after complete unloading following an elastic–plastic contact with a rigid conical indenter. The sides of the impression are not straight but slightly curved because the material near the tip of the indenter has not recovered to the extent expected upon theoretical grounds (Courtesy CSIRO and after reference 5).

The conclusion is thus formed that the shape of the unloading curves observed experimentally is due to the curved sides of the residual impression, the curvature being more pronounced in materials with a lower value of E/H. Further, the observation that the depth of the residual impression is greater than that expected on theoretical grounds leads one to conclude that the curvature is not arising from more than expected elastic recovery of the flanks of the impression, but less than expected elastic recovery of the material near the tip of the indenter and is possibly due to densification within the relatively open structure of the fused silica material. These results demonstrate that the shape of the residual impression, especially for a highly elastic solid, does not correspond to that assumed by the theoretical treatments used in the conventional analysis of this type of data. While the initial part of the unloading curves appears to be not so much affected, it is still of some significance since it is the slope of the tangent to the initial unloading response, extrapolated to the depth axis, that is used in the analysis calculations. Any variation in the estimate of this slope, due to a large leverage effect, has a significant effect on the estimation of the plastic depth h_p used in the analysis with values of E increasing and H decreasing with an increasing number of unloading points used in the fitting procedure. Mathematically, the significance of the curved shape of the residual impression means that the exponent in a power law fitting to the unloading data for a conical indenter is not 2 as would be expected from the elastic equations of contact but rather, in the range 1.2 – 1.6[7].

References

1. J.S. Field and M.V. Swain, "A simple predictive model for spherical indentation," J. Mater. Res. 8 2, 1993, pp. 297–306.
2. W.C. Oliver and G.M. Pharr, "An improved technique for determining hardness and elastic modulus using load and displacement sensing indentation experiments," J. Mater. Res. 7 4, 1992, pp. 1564–1583.
3. R.B. King, "Elastic analysis of some punch problems for a layered medium," Int. J. Solids Structures, 23 12, 1987, pp. 1657–1664.
4. W.A. Caw, "The elastic behaviour of a sharp obtuse wedge impressed on a plane," J. Sci. Instr., J. Physics E, 2 2, 1969, pp. 73–78.
5. A.C. Fischer-Cripps, "Elastic recovery and reloading of hardness impressions with a conical indenter," Mat. Res. Soc. Symp. Proc. 750, 2003, pp. Y6.9.1–Y.6.9.6.
6. N.A. Stilwell and D. Tabor, "Elastic recovery of conical indentations," Phys. Proc. Soc. 78 2, 1961, pp. 169–179.
7. G.M. Pharr and A. Bolshakov, "Understanding indentation unloading curves," J. Mater. Res. 17 10, 2002, pp. 2660–2671.

Chapter 6
Scaling Relationships in Nanoindentation

6.1 Scaling Relationships in Nanoindentation

An interesting fundamental approach to the analysis of load-displacement data is provided by dimensional analysis.[1–9] Consider the indentation of an elastic–plastic specimen with a rigid conical indenter. The mechanical properties of the specimen can be approximated by a uniaxial stress–strain response given by Eqs. 4.6a and 4.6b, here repeated for convenience:

$$\begin{aligned} \sigma &= E\varepsilon \qquad \varepsilon \leq Y/E \\ \sigma &= K\varepsilon^{x} \qquad \varepsilon \geq Y/E \end{aligned} \tag{6.1a}$$

where σ is the applied stress and ε is the resulting strain and K is equal to:

$$K = Y\left[\frac{E}{Y}\right]^{x} \tag{6.1b}$$

In Eq. 6.1b, x is the strain-hardening exponent of the material. For $x = 0$, the solid is elastic perfectly-plastic.

During the loading portion of an indentation test, the quantities of interest are the radius of the circle of contact a, the depth of the contact h_p, and the load, from which the hardness can be calculated. Consider the two key experimental variables of interest, the indenter load P and total penetration h. For a given value of penetration depth, the required load must be a function of E, ν, Y, n, and h and, of course, the angle of the indenter α. We may ask ourselves, of these variables, which are the governing parameters? That is, which of these parameters set the dimensions of the others? Obviously, ν and n have no dimensions, and neither does α. Evidently, the dimension of depth is important, and we are faced with a choice of selecting E or Y, since one sets the dimensions of the other. Let us therefore select E and h as the governing parameters for dimensional analysis. The dimensions of these set the dimensions of the others:

$$[Y] = [E]$$
$$[P] = [E][h]^2 \tag{6.1c}$$

The contact depth h_p, which eventually provides the value of hardness, is desired to be expressed as a function g of these parameters. Similarly, the load P is to be expressed as a function f_L (for "loading") of these same parameters:

$$h_c = g(E, \nu, Y, n, h, \alpha)$$
$$P = f_L(E, \nu, Y, n, h, \alpha) \tag{6.1d}$$

The Buckingham Pi theorem in dimensional analysis yields:

$$\Pi_\alpha = \Pi_\alpha(\Pi_1, \nu, n, \alpha) \tag{6.1e}$$

which can be written:

$$P = Eh^2\Pi_\alpha\left(\frac{Y}{E}, \nu, n, \alpha\right) \tag{6.1f}$$

where

$$\Pi_\alpha(\Pi_1, \nu, n, \alpha) = \frac{P}{Eh^2}$$
$$\Pi_1 = \frac{Y}{E} \tag{6.1g}$$

Since Π_α, Y/E, ν, n, α are all dimensionless, we can see that the force F is proportional to the square of the indentation depth h. A similar treatment provides information about the contact depth h_p:

$$h_p = h\Pi_\beta\left(\frac{Y}{E}, \nu, n, \alpha\right) \tag{6.1h}$$

where it is seen that the contact depth h_c is directly proportional to the total penetration depth h, the constant of proportionality being dependent on Y/E and n for a given indenter angle α.

During unloading, the load P also depends, in addition to the above parameters, on the maximum depth of penetration:

$$P = f_L(E, \nu, Y, n, h, h_t, \alpha) \tag{6.1i}$$

Dimensional analysis yields:

$$P = Eh^2\Pi_\gamma\left(\frac{Y}{E}, \frac{h}{h_t}, \nu, n, \alpha\right) \tag{6.1j}$$

Equation 6.1j shows that the load P now depends on h^2 and the ratio h/h_t. Taking the derivative of Eq. 6.1j with respect to h, and evaluating this at $h = h_t$, the form of the slope of the initial unloading becomes:

$$\frac{dP}{dh} = Eh_t\Pi_\delta\left(\frac{Y}{E}, \nu, n, \alpha\right) \tag{6.1j}$$

where it is shown that the slope of the initial unloading is proportional to h_t (for a constant value of Y/E, ν, n and α). The residual depth hr at P = 0 is found from Eq. 6.1j:

$$h_r = h_t\Pi_\phi\left(\frac{Y}{E}, \nu, n, \alpha\right) \tag{6.1k}$$

The total work done during the loading part of the indentation cycle is found from:

$$\begin{aligned} W_T &= \int_0^{h_t} F dh \\ &= \frac{Eh_t^3}{3}\Pi_\alpha\left(\frac{Y}{E}, \nu, n, \alpha\right) \end{aligned} \tag{6.1l}$$

During unloading, the work done by the solid on the indenter is:

$$\begin{aligned} W_T &= \int_{h_r}^{h_t} F dh \\ &= \frac{Eh_t^3}{3}\Pi_U\left(\frac{Y}{E}, \nu, n, \alpha\right) \end{aligned} \tag{6.1m}$$

The ratio of irreversible work, or energy dissipated within the solid, to the total work is independent of h_t and becomes:

$$\frac{W_T - W_U}{W_T} = \Pi_W\left(\frac{Y}{E}, \nu, n, \alpha\right) \tag{6.1n}$$

Finite element analysis can be used to test the relationships of Eqs. 6.1f and 6.1h and also the form of the dimensionless functions $\Pi_{\alpha,\beta,\gamma,\phi}$. These scaling relationships are important because they allow us to test the relationships between experimental variables within a theoretical framework. For example, comparison of finite element results shows that the quantities h_r/h_t (from Eq. 6.1k) and Π_W (from Eq. 6.1n) are linearly dependent.

As another example, consideration of the ratio h_p/h (see Eq. 6.1h) shows that it can be greater than or less than one, corresponding to piling-up and sinking-in, respectively. For large values of Y/E, sinking-in occurs for $n > 0$. For small values of Y/E, both piling-up and sinking-in can occur depending on the value of n. When Eq. (3.2.3.1e) is combined with Eq. 3.2.3c, we obtain:

$$h_p = h_t - P\varepsilon \frac{dh}{dP} \tag{6.1o}$$

where in Eq. 6.1o, ε is the intercept factor in Table 1.1. Comparison with the results of finite element analysis shows that the multiple point unload method (the "Oliver and Pharr" method) is valid only for the case of sinking-in (large values of Y/E).

Since the radius of the circle of contact for a conical indenter is given by Eq. 1.2o, we can write, using Eqs. 6.1h and 6.1j:

$$\frac{dP}{dh} = Ea \cot\alpha \left[\frac{\Pi_\alpha (Y/E, \nu, n, \alpha)}{\Pi_\beta (Y/E, \nu, n, \alpha)} \right] \tag{6.1p}$$

Finite element analysis shows that the quantity dP/dh divided by the product Ea for particular values of a is independent of Y/E and n and is approximately equal to 2 in accordance with Eq. 3.2.3.1h. This result confirms that the elastic modulus can be calculated from the initial unloading slope provided the contact radius a is known.

As another application of the scaling relationships, with H being given by Eq. 3.2.3d, the ratio H/Y becomes, by Eqs. 6.1f and 6.1h:

$$\frac{H}{Y} = \frac{\cot^2\alpha}{\pi} \left[\frac{\Pi_\alpha (Y/E, \nu, n, \alpha)}{Y/E\, \Pi_\beta{}^2 (Y/E, \nu, n, \alpha)} \right] \tag{6.1q}$$

which implies that the hardness H is independent of depth h and that the constraint factor C depends upon the ratio Y/E. Finite element analysis shows that the constraint factor varies between 1.7 and 2.8 increasing with decreasing Y/E.

Given the mechanical properties of a material, the essential features of a nanoindentation loading and unloading response can be predicted by the scaling relationships given above. However, given a load-displacement curve, say from an experiment, there is not always a unique solution to the associated mechanical properties. While a value for modulus E can be found from the loading curves (Eq. 6.1f), values for Y and n cannot be uniquely extracted from the unloading curves.

Scaling relationships allow the estimation of material parameters even when the geometry of the indenter is not ideal. For example, from Eq. 4.9a, the loading with a pyramidal indenter (of equivalent cone angle 70.3°) with a rounded tip of radius R should be identical with that of a spherical indenter for $h_t/R < 0.073$ and only approach that of a sharp indenter when $h_t/R >> 0.073$. By fitting a second-order polynomial to the load-displacement curve, the coefficients can be used to determine the yield stress Y and the tip radius R if the elastic modulus E of the specimen is known.[3]

The method of dimensional analysis can be extended to cases involving non-linear solids, such as those that follow a power-law creep response.[9–11] In such a material, the mechanical response of the material depends upon the rate of application of strain:

$$\sigma = K\dot{\varepsilon}^{m} \tag{6.1r}$$

where K and m are material constants. Introducing the dimension of time into the problem yields the following dimensional relationship for hardness H:

$$H = K\left(\frac{\dot{h}}{h}\right)^{m} \frac{\Pi_{\alpha}^{c}}{\Pi_{\beta}^{c}} \tag{6.1s}$$

where Π in Eq. 6.1s are dimensionless functions of both K and m. The importance of Eq. 6.1s is that it demonstrates the strain rate dependence on hardness observed in experimental work[12]. Equation 6.1s can be used to predict the load-displacement response of a power-law creeping solid under different conditions of application of load. For example, for a constant displacement rate, it is shown that the load is no longer proportional to h^2 but proportional to h^{2-m} and the hardness decreases with increasing indentation depth. For a constant load rate, it is shown that the hardness increases with increasing load — the latter two conditions leading to an observed indentation size effect. The measured hardness in an indentation test is observed to reach a steady-state value when the load rate, divided by the load, is held constant during the test.

The generality of this method of dimensional analysis has the potential to yield new information about other mechanisms of deformation in indentation experiments.

References

1. C.-M. Cheng and Y.-T. Cheng, "On the initial unloading slope in indentation of elastic–plastic solids by an indenter with an axisymmetric smooth profile," Appl. Phys. Lett. 71 18, 1997, pp. 2623–2625.
2. Y.-T. Cheng and C.-M. Cheng, "Relationships between hardness, elastic modulus, and the work of indentation," Appl. Phys. Lett. 73 5, 1997, pp. 614–616.
3. Y.-T. Cheng and C.-M. Cheng, "Further analysis of indentation loading curves: Effects of tip rounding on mechanical property measurements," J. Mater. Res. 13 4, 1998, pp. 1059–1064.

4. Y.-T. Cheng and C.-M. Cheng, "Effects of 'sinking in' and 'piling up' on the estimating the contact area under load in indentation," Phil. Mag. Lett. 78 2, 1998, pp. 115–120.
5. Y.-T. Cheng and C.-M. Cheng, "Scaling approach to conical indentation in elastic–plastic solids with work-hardening," J. App. Phys. 84 3, 1998, pp. 1284–1291.
6. Y.-T. Cheng and C.-M. Cheng, "Analysis of indentation loading curves obtained using conical indenters," Phil. Mag. Lett. 77 1, 1998, pp. 39–47.
7. Y.-T. Cheng and C.-M. Cheng, "Scaling relationships in conical indentation of elastic-perfectly plastic solids," Int. J. Solids Structures, 36, 1999, pp. 1231–1243.
8. Y.-T. Cheng and C.-M. Cheng, "Can stress–strain relationships be obtained from indentation curves using conical and pyramidal indenters?," J. Mater. Res. 14 9, 1999, pp. 3493–3496.
9. Y.-T. Cheng and C.-M. Cheng, "What is indentation hardness?," Surf. Coat. Tech. 133-134, 2000, pp. 417–424.
10. R. Hill, "Similarity of creep indentation tests," Proc. Roy. Soc. A436, 1992, pp. 617–630.
11. Y.-T. Cheng and C.-M. Cheng, "Scaling relationships in indentation of power-law creep solids using self-similar indenters," Phil. Mag. Lett. 81 1, 2001, pp. 9–16.
12. M.J. Mayo and W.D. Nix, "Measuring and Understanding Strain Rate Sensitive Deformation with the Nanoindenter," Proc. 8th int. Conf. On Strength of Metals and Alloys, Tampere, Finland, P.O. Kettunen, T.K. Lepisto and M.E. Lehtonen, eds. Pergammon Press, Oxford, 1988, pp. 1415–1420.

Chapter 7
Time-Dependent Nanoindentation

7.1 Introduction

In general, materials can resist deformation in a solid-like or viscous-like manner. Solid-like materials store energy under deformation, and upon removal of stress, returns to its original state. Viscous materials dissipate energy during deformation and upon removal of stress, remains in its deformed state. Materials with combined solid-like and viscous-like properties are said to be viscoelastic. Nanoindentation can be used to quantitatively determine the viscoelastic properties of materials. In one method, a small oscillatory force or displacement is imparted to the indenter. The resulting load and displacement signals provide a method whereby the elastic and viscous components of the specimen response can be calculated. In another method, the load or displacement is held at a fixed value and the change in displacement (creep) or load (relaxation) recorded over a period of time. Application of an appropriate mechanical model can yield values for the elastic and viscous properties of the specimen.

In some respects, the oscillatory mode of operation in dynamic nanoindentation testing is similar to that which occurs in an atomic force microscope (AFM), but there are several practical differences. In an AFM, the spring stiffness of the cantilever that supports the probe is made very compliant so as to enhance the force resolution of the instrument and to avoid unintentionally damaging the specimen surface. The AFM is primarily an imaging device. For a nanoindentation instrument, the stiffness of the supporting springs for the indenter is high so as to provide enough load to enable the mechanical properties (rather than the surface topography) to be measured. However, several studies of nanoindentation using an AFM have been reported in the literature[1,2]. Syed Asif, Colton, and Wahl[3,4] have coupled a dynamic mode of testing with a Hysitron TriboScope®, a traditional nanoindentation instrument that utilizes the piezo-controlled specimen positioner of an AFM, which they refer to as a "hybrid nanoindenter." The resulting dynamic mode of testing offers the ability to measure surface forces, surface energies, interaction stiffness prior to contact, and a quantitative imaging technique for mechanical property mapping at the nano scale.

7.2 Dynamic Indentation Testing

7.2.1 Single Frequency Dynamic Analysis

The solid-like or viscous-like nature of materials is quantified by the stress response to an applied strain. In an ideal solid-like material, the stress response will be in phase with the strain, i.e., the largest deformation coincides with the largest stress, and the constant of proportionality between the two is referred to as the elastic modulus E. In dynamic tests, we are usually concerned with viscoelastic behavior and it is the shear modulus G that is often of most interest since experimental work is usually designed to apply shear stresses and strains to the specimen rather than uniform tension or compression. Both E and G are material properties which quantify the elastic properties of a material and hence their ability to store elastic strain energy under load. For this reason, E or G (depending on the nature of the loading) is referred to as the storage modulus and given the symbol G'. In an ideal viscous material, there is no elastic deformation and the stress response will be 90° out of phase with the applied strain. In this case, the resulting stress in the material is proportional to the rate of applied strain, the constant of proportionality is referred to as the loss modulus G" and is directly related to the viscosity of the specimen. Energy expended in deforming the material in this case is lost as heat. For viscoelastic solids, the combined elastic and viscous response is represented by the complex modulus defined as:

$$G^{*} = G' + iG'' \tag{7.2.1a}$$

Consider a mass supported by a spring of stiffness k and dashpot of damping coefficient λ as shown in Figs. 7.1 (a) and (b). The arrangement shown in Fig. 7.1 (a) is called a Voigt element, and that in Fig. 7.1 (b) is called a Maxwell element. If a force p(t) is applied to the mass so as to cause the mass to oscillate, then, in the general, there is a phase difference ϕ between magnitudes of the force and displacement. Mathematically, this is expressed as:

$$\begin{aligned} P(t) &= P_0 e^{i(\omega t + \phi)} \\ h(t) &= h_0 e^{i(\omega t)} \end{aligned} \tag{7.2.1b}$$

Taking the first and second derivatives with respect to time gives the velocity and acceleration of the mass:

$$\begin{aligned} \frac{dh}{dt} &= i\omega h \\ \frac{d^2h}{dt^2} &= -\omega^2 h \end{aligned} \tag{7.2.1c}$$

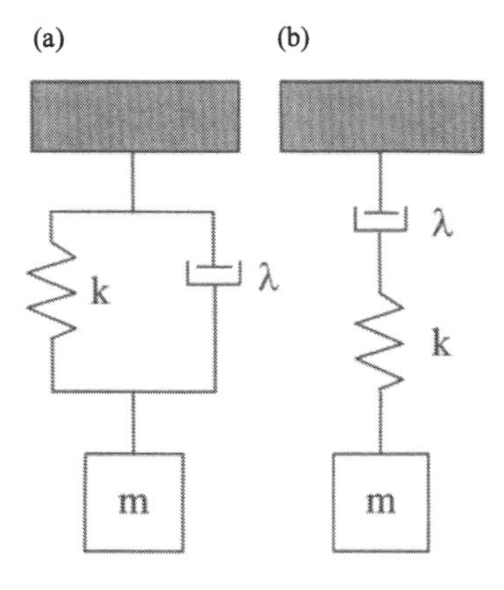

Fig. 7.1 Dynamic mechanical model of a mass supported by a spring of stiffness k and dashpot of damping coefficient λ. (a) Voigt model (b) Maxwell model. A sinusoidal force p(t) is applied to the mass resulting in a motion h(t).

Now, for the case of (a) in Fig. 7.1, since the spring and dashpot elements are in parallel, then the force is given by the sum of the components thus:

$$\begin{aligned} p &= kh + \lambda\frac{dh}{dt} - m\frac{d^2h}{dt^2} \\ &= kh + \lambda i\omega h - m\omega^2 h \\ &= \left(k - m\omega^2 + i\omega\lambda\right)h \end{aligned} \tag{7.2.1d}$$

where both p and h are functions of time. The bracketed term in Eq. 7.2.1d is called the transfer function. Terms within the transfer function have physical significance. For the system shown in Fig. 7.1, the storage modulus G' and the loss modulus G" are found from the real and imaginary components of the transfer function:

$$\begin{aligned} G' &= k \\ G'' &= i\omega\lambda \end{aligned} \tag{7.2.1e}$$

Where the loading geometry is more complicated than that shown in Fig. 7.1, such as in an indentation test, the storage and loss moduli can be obtained from the transfer function if the relevant geometrical loading parameters are known. For example, in an indentation test, the spring constant k in the above equations is the contact stiffness S and is related to the storage modulus G', which in this case, is the combined elastic modulus of the indenter and specimen E^* by:

$$E^* = S\frac{\sqrt{\pi}}{2\sqrt{A}} \tag{7.2.1f}$$

In principle, the area of contact A can be determined from the plastic depth h_p according to the geometry of the indenter (see Table 1.1) in conjunction with Eq. 3.2.4.1d.

If S represents the contact stiffness, then the magnitude of the transfer function for Eq. 7.2.1d is given by:

$$\begin{aligned} |TF| &= \left|\frac{P_o}{h_o}\right| \\ &= \sqrt{\left(S - m\omega^2\right)^2 + \omega^2\lambda^2} \end{aligned} \tag{7.2.1g}$$

and the phase angle between the force and displacement is found from:

$$\tan\phi = \frac{\omega\lambda}{S - m\omega^2} \tag{7.2.1h}$$

In a nanoindentation experiment, the magnitudes P_o and h_o and the phase angle $\tan\phi$ are measured. The mass term can be found by oscillating the indenter in air (i.e. $S = 0$ and $\lambda = 0$) and sweeping through a range of frequencies whereupon m is found from a plot of the magnitude of the transfer function against ω^2 and represents the mass of the oscillating indenter and shaft. For contact with a viscoelastic specimen, the real and imaginary parts of the transfer function are determined from experimental readings of P_o, h_o and ϕ as:

$$\begin{aligned} TF_{re} &= \left|\frac{P_o}{h_o}\right|\cos\phi = S - m\omega^2 \\ TF_{im} &= \left|\frac{P_o}{h_o}\right|\sin\phi = \omega\lambda \end{aligned} \tag{7.2.1i}$$

from which S may be determined and hence E^* using Eq. 7.2.1f.

The systems shown in Fig. 7.1 (a) and (b) are just two simple representations of specimen behavior. In practice, there are additional mechanical elements involved in the deformation that depend upon the nature of the instrument being used to perform the test. For example, in an indentation test, the stiffness of the indenter shaft support springs k_s must be taken into consideration as shown in Fig. 7.2. As before, a small, modulated force p(t) is applied with a frequency ω and amplitude P_o according to Eq. 7.2.1b.

An analysis of the model shown in Fig. 7.2 gives the following transfer function:

$$p = \left(S + k_s - m\omega^2 + i\omega\lambda\right)h \tag{7.2.1j}$$

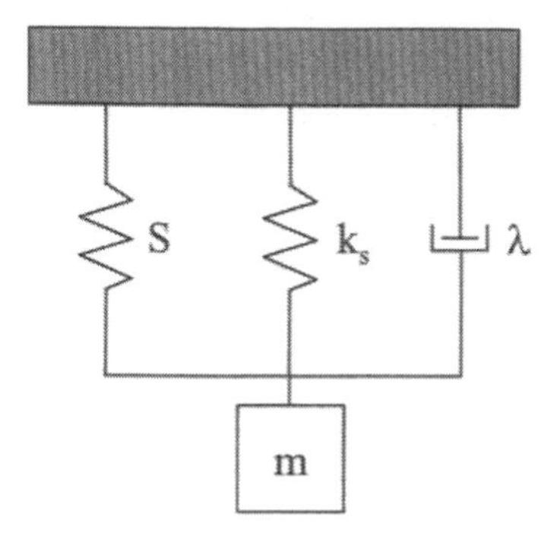

Fig. 7.2 Dynamic 3-element mechanical model of a mass supported by springs of stiffness S and k_s and dashpot of damping coefficient λ.

The magnitude of the transfer function is:

$$\left|\frac{P_o}{h_o}\right| = \sqrt{\left(S + k_s - m\omega^2\right)^2 + \omega^2\lambda^2} \tag{7.2.1k}$$

and the phase difference φ is given by:

$$\tan\phi = \frac{\omega\lambda}{S + k_s - m\omega^2} \tag{7.2.1l}$$

In Eqs. 7.2.1j, 7.2.1k and 7.2.1l, k_s is the stiffness of the indenter shaft support springs, P_o is the magnitude of the oscillatory load, h_o is the magnitude of the displacement oscillation, ω is the frequency of the oscillation, m is the mass of the oscillating components, λ is a damping coefficient associated with the contact and S is the contact stiffness dP/dh. The real and imaginary parts of the transfer function are thus written:

$$\begin{aligned} TF_{re} &= S + k_s - m\omega^2 \\ TF_{im} &= i\omega\lambda \end{aligned} \tag{7.2.1m}$$

As before, values for k_s and m can be determined by fitting to the transfer function obtained by oscillating the indenter in air.

Conventionally, elements of the measurement system and the tip-specimen interaction are accommodated by the addition of spring and dashpot elements as appropriate. A popular implementation of the dynamic indentation testing technique[5] uses the model shown in Fig. 7.3 where the stiffness S determined from measurements of P_o and h_o at a frequency ω to determine the hardness as a function of indentation depth during an indentation test. The transfer function is expressed as:

$$\left|\frac{P_o}{h_o}\right| = \sqrt{\left(\left(\frac{1}{S}+C_f\right)^{-1} + k_s - m\omega^2\right)^2 + \omega^2\lambda^2} \tag{7.2.1n}$$

where C_f is the compliance of the load frame. The phase difference ϕ between the load and displacement is given by:

$$\tan\phi = \frac{\omega\lambda}{\left(\frac{1}{S}+C_f\right)^{-1} + k_s - m\omega^2} \tag{7.2.1o}$$

Values of k_s and λ are determined from readings obtained with the indenter clear of the specimen surface ($S = 0$) and sweeping over the frequency range of interest.

When the specimen to be measured is viscoelastic, there is an additional damping term λ_s associated with the tip–specimen interaction. For tests done over a range of frequencies, a plot of stiffness S versus indenter displacement is proportional to the storage modulus of the material, and a plot of damping coefficient times the frequency $\lambda_s\omega$ versus displacement is proportional to the loss modulus[6] of the material at the contact interaction.

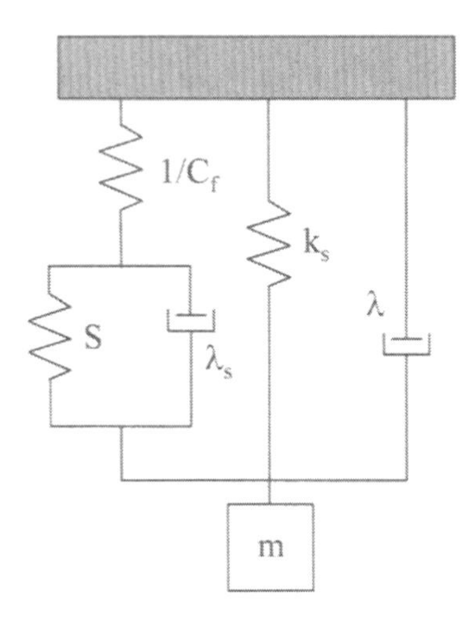

Fig. 7.3 Dynamic mechanical model of a nanoindentation instrument that includes the compliance of the load frame and the tip/specimen interaction. k_s is the stiffness of the indenter shaft support springs, λ is a damping coefficient representing the instrument response, S is the stiffness of the contact, C_f is the compliance of the load frame, λ_s is a damping coefficient of the contact, and m is the mass of the indenter and shaft.

If E^* is known, then the contact area A can be calculated from Eq. 3.2.1c (with S = dP/dh) and hence hardness as a function of indentation depth can be obtained quite readily. Alternately, the area of contact can be found from h_p calculated from Eq. 3.2.4.1d allowing both hardness and modulus to be determined.

Burnham, Baker, and Pollock[7] have prepared a universal mechanical model that accounts for the characteristics of various types of indentation instruments and that of the tip-specimen interaction. In this work, the elements of the mechanical model are placed in series as shown in Fig. 7.4. In the improved model of Fig. 7.4, the velocities on all sides of the elements are accounted for in the expressions for amplitude and phase of the motion.

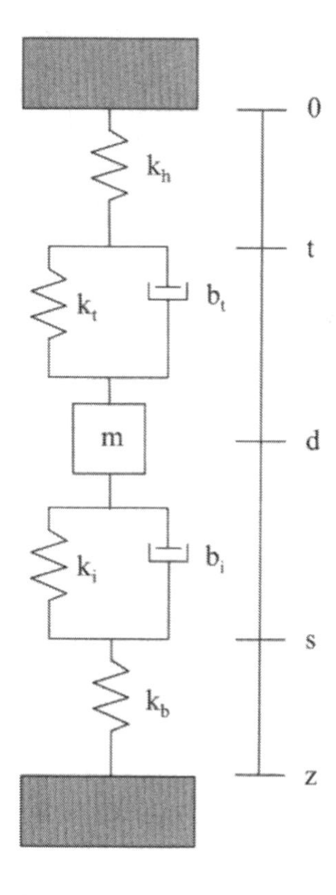

Fig. 7.4 Universal dynamic mechanical model of a nanoindentation instrument. k_t is the stiffness of the force actuator with damping b_t. The tip–specimen interaction has a stiffness k_i (dP/dh) and damping b_i. k_h and k_b are the stiffnesses of the head and base of the load frame, respectively.

The resulting expressions take the form:

$$\frac{d-t}{p}=\sqrt{\left[\left(1+\frac{k_i}{k_b}\right)^2+\left(\frac{\omega b_i}{k_b}\right)^2\right]\left[K^2+(\omega B)^2\right]^{-1}} \tag{7.2.1p}$$

where d and t are the points labeled on Fig. 7.4.

The phase difference is given by:

$$\tan\phi=\omega\left[\frac{b_i}{k_b}K-\left(1+\frac{k_i}{k_b}\right)B\right]\left[\left(1+\frac{k_i}{k_b}\right)K+\omega^2\frac{b_i}{k_b}B\right]^{-1} \tag{7.2.1q}$$

where K and B are an effective stiffness and damping, respectively, for the system. In the limit of a very stiff machine, where k_t, k_i, b_t, b_i are very much less than k_h and k_b, Eqs. 7.2.1p and 7.2.1q reduce to 7.2.1n and 7.2.1o.

The complexity of mechanical models such as those shown in Figs. 7.3 and 7.4 makes analysis of dynamic indentation test results difficult. An alternative is to model the response of the specimen only, and to account for any instrument characteristics by virtue of a test on a standard reference specimen.

Linear viscoelastic solids are generally best represented by a standard three-parameter Voigt model as shown in Fig. 7.5. In this model, an application of a step strain or displacement results in an immediate high value of stress that relaxes over time to a steady-state value. A step increase in stress results in an immediate displacement that increases to a steady-state value over time.

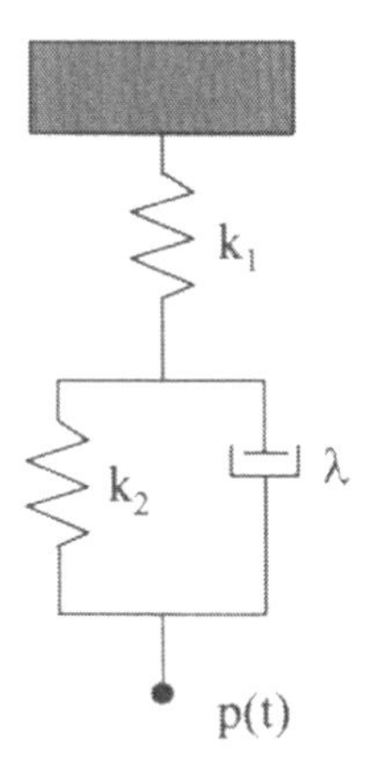

Fig. 7.5 Standard Voigt 3-element mechanical model of a structural material consisting of springs of stiffness k_1 and k_2 and dashpot of damping coefficient λ. A sinusoidal force p(t) is applied.

In Fig. 7.5, the mass element has not been included since the model represents the specimen only in an indentation test and not the oscillating components of the instrument (i.e. indenter shaft) with a mass term. These elements are not included since, as will be shown later, they can be accounted for by reference tests on a standard specimen. For a step increase in load to $p = P_o$, the relationship between the resulting displacement and time (i.e. steady creep) is given by:

$$h = P_o\left(\frac{1}{k_1} + \frac{1}{k_2}\left(1 - e^{-t\frac{k_2}{\lambda}}\right)\right) \tag{7.2.1r}$$

The resulting transfer function for oscillatory motion has the form:

$$p = h\left(\frac{1}{k_2 + i\omega\lambda} + \frac{1}{k_1}\right)^{-1} \tag{7.2.1s}$$

Separation of Eq. 7.2.1s into real and imaginary parts yields:

$$p = h\left[\frac{\left(k_1 k_2 (k_1 + k_2) + k_1 \lambda^2 \omega^2\right) + i\omega\lambda k_1^{\ 2}}{(k_1 + k_2)^2 + \omega^2\lambda^2}\right] \tag{7.2.1t}$$

The real and imaginary parts of the transfer function are thus given in terms of the spring k and damping λ terms:

$$\begin{aligned} TF_{re} &= \frac{|P_o|}{|h_o|}\cos\phi = \left[\frac{k_1 k_2 (k_1 + k_2) + k_1 \lambda^2 \omega^2}{(k_1 + k_2)^2 + \omega^2 \lambda^2}\right] \\ TF_{im} &= \frac{|P_o|}{|h_o|}\sin\phi = \frac{\lambda\omega k_1^{\ 2}}{(k_1 + k_2)^2 + \omega^2\lambda^2} \end{aligned} \tag{7.2.1u}$$

In the quasi-static case ($\omega \to 0$), TF_{re} reduces to the series addition of k_1 and k_2 as would be expected intuitively. At very high frequencies, TF_{re} reduces to k_1. As the damping term λ becomes larger at a particular value of frequency ω, TF_{re} and the effective contact stiffness become dominated by the value of k_1.

The square-bracketed term in Eq. 7.2.1t is the transfer function for the contact. The real part of the transfer function represents the contact stiffness S since the model in this case represents the specimen only (i.e. no indenter mass or support spring stiffness terms are included here). The method in which these instrument terms are accounted for will be described in Section 7.2.2. The combined modulus E^* is thus effectively frequency dependent and is expressed, by virtue of Eq. 7.2.1f, as:

$$\begin{aligned} E^* &= \frac{\sqrt{\pi}}{2\sqrt{A}} \frac{|P_o|}{|h_o|} \cos\phi \\ &= \frac{\sqrt{\pi}}{2\sqrt{A}} TF_{re} \end{aligned} \tag{7.2.1v}$$

Equation 7.2.1u consists of two equations and three unknowns and so it is evident that a fitting procedure is required to determine the individual values of k_1, k_2 and λ from the measured transfer functions. However, for most cases, a combined value for E^* from Eq. 7.2.1v is all that is required. The area of contact is found from the plastic depth h_p according to the geometry of the indenter in conjunction with Eq. 3.2.4.1d and the measured values of S, P_t and h_t.

Equivalent interpretations are required to extract the viscosity η of the specimen material. Eq. 7.2.1e shows that for the model shown in Fig. 7.1 (a), the damping term λ is immediately obtained from the loss modulus G" where for the model shown in Fig. 7.5, Eq. 7.2.1u is appropriate. In the case of indentation loading, the situation is more complicated since the area of contact changes upon loading and unloading and so the nature of the viscous part of the contact depends upon the loading history. It is customary to equate the quantity $\omega\lambda$ with the loss modulus G" (see Eq. 7.2.1m) but it is the viscosity that is of real physical interest. For the case of step loading of a linear viscous creeping solid with a conical indenter, the relationship between the load and depth of penetration has the form[8]:

$$P = \left(\frac{8}{3}\eta\right)\left[\frac{2}{\pi} h \tan\alpha\right]\frac{dh}{dt} \tag{7.2.1w}$$

If we make the substitution in 7.2.1c, then we may arrange Eq. 7.2.1w thus:

$$P = i\left(\omega\frac{8}{3}\eta\right)\frac{2}{\pi} h^2 \tan\alpha \tag{7.2.1x}$$

where the similarity to the corresponding elastic equation in Table 1.1 should be immediately apparent. The bracketed term in Eq. 7.2.1w therefore corresponds to a loss modulus expressed as:

$$\begin{aligned} E'' &= \frac{\sqrt{\pi}}{2\sqrt{A}} \frac{|P_o|}{|h_o|} \sin\phi \\ &= \frac{\sqrt{\pi}}{2\sqrt{A}} TF_{im} \end{aligned} \tag{7.2.1y}$$

where TF_{im} is determined from experimental readings. The viscosity η is thus obtained by dividing E" by $8\omega/3$. A similar analysis for the case of a spherical indenter gives:

$$P = j(2\omega\eta)\frac{4}{3}h^{3/2}R^{1/2} \qquad (7.2.1z)$$

One must realize that Eq. 7.2.1w applies only to the loading part of the deformation and to extend the result to a sinusoidal oscillatory motion (loading and unloading) may not be entirely appropriate. Further, in many contact situations, the relationship between P and h is non-linear, the function depending upon the shape of the indenter. The presence of plastic deformation during the contact also affects the validity of the zero frequency value of E^* calculated using the above procedure. In practice, the analysis given above is appropriate for small amplitude oscillations only.

7.2.2 Multiple Frequency Dynamic Analysis

The above techniques can be further extended into multiple frequency testing in combination with Fourier analysis. This is possible when it is recognized that the transfer function between load and displacement is given by the ratio of the Fourier transform of the component signals.

$$TF = \frac{F(P(t))}{F(h(t))} \qquad (7.2.2a)$$

The significance of this is that the Fourier transforms of the load and displacement signals can be calculated from experimental data. The Fourier transforms are complex quantities and are expressed as a function of frequency. The transfer function TF is a complex quantity, the real part leads to the storage modulus and the imaginary part gives the loss modulus. If a force or displacement signal containing multiple frequencies is applied, then the transfer function at each component frequency can readily be calculated. Physical quantities can then be obtained using the appropriate contact equations.

There is a practical requirement to obtain a suitable input force signal that represents the desired frequency components. Simple addition of p(t) of different frequencies at a constant amplitude results in a response that approximates a step input in the limit of an infinite number of components. It is more practical therefore to combine a series of frequency components that have a random distribution of phase so that the resulting input signal remains within the capabilities of the measurement instrument. Fig. 7.6 shows a typical shape of p(t) for a random distribution of phases for a number of component frequencies between zero and 100 Hz.

For nanoindentation loading, a simple example of an appropriate modeling equation for the tip-specimen interaction is given by Eq. 7.2.1d whereupon the real part of the transfer function provides a value for the storage modulus G' and the imaginary part provides a value for the loss modulus G" according to Eq. 7.2.1e. That such a simple model can be used to extract physical quantities of interest (e.g. Eq. 7.2.1f) is possible due to the cancellation of the instrument response by a procedure called equalization.

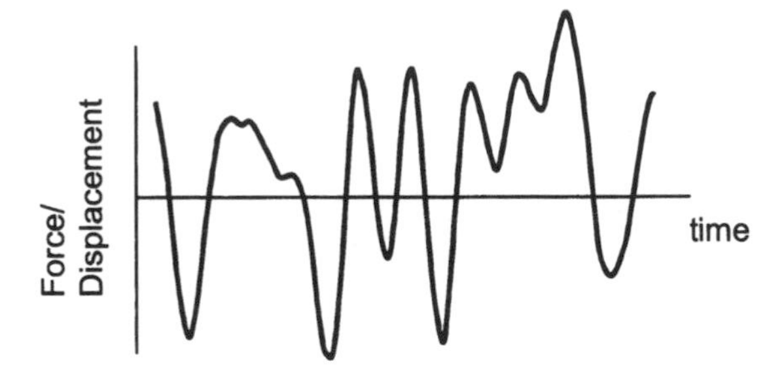

Fig. 7.6 Psuedo-random load signal applied to the indenter. The signal is constructed from sinusoidal components of a range of frequencies with the same amplitude, but a random phase.

In general, the transfer function as measured by the instrument contains viscous and elastic contributions from both the instrument and the specimen. More formally, the force and displacement signals obtained when testing a specimen contain a convolution of the force and displacement responses in the time domain of the specimen and the instrument. Previously, it was suggested that the instrument component could be obtained by oscillating the indenter shaft in air and this is done in some commercial instruments. An alternative approach is to apply an oscillatory motion to a reference spring or a dashpot whose stiffness or damping factor is already known. Eq. 7.2.2a shows that the transfer function is expressed as the ratio of the Fourier transforms of the force and displacement signals and is thus in the frequency domain. Convolutions in the time domain become multiplications in the frequency domain, hence, we form the equalised transfer function as:

$$TF_{Eq} = \frac{\left(TF_{Specimen} TF_{Instrument}\right)}{\left(TF_{Instrument} TF_{Reference}\right)} \tag{7.2.2b}$$

where the desired quantity to be determined is $TF_{Specimen}$. The bracketed terms in Eq. 7.2.2b are the transfer functions obtained directly from experimental data obtained by tests on the actual specimen and the reference specimen. $TF_{Reference}$ on its own is the known transfer function of the reference, which in the case of an elastic spring, is its stiffness as measured independently.

The transfer functions above are complex quantities, having both magnitude and phase. Thus, the transfer function of the specimen only is obtained from:

$$TF_{Specimen} = TF_{Eq} TF_{Reference} \tag{7.2.2c}$$

Often it is desirable to select a reference specimen with well known, frequency independent quantities (at least over the frequency range of interest). A simple linear spring or Newtonian fluid are ideal choices. Physical quantities of interest are then derived from the modeling equations in Section 7.2.1.

7.3 Creep

The methods of analysis of nanoindentation load-displacement curves discussed in Chapter 3 assume that yield (or irreversible deformation) occurs instantaneously as a result of some physical mechanism such as shear-driven slip or plastic flow. It was assumed that the material behaved in an elastic–plastic manner and did not exhibit any time-dependent behavior or load rate dependence. In many materials, time-dependent creep within the specimen can occur under indentation loading and manifests itself as a change of the indentation depth under a constant applied load.

In a nanoindentation test, creep and plastic deformation in the conventional sense, i.e., that occurring due to shear-driven slip for example, should be regarded separately. Plasticity, in the sense of yield or hardness, is conveniently thought of as being an instantaneous event (although in practice, it can take time for yield processes to occur). In contrast, creep can occur over time in an otherwise elastic deformation as a result of the diffusion and motion of atoms or movement of dislocations in the indentation stress field, the extent of which depends very much on the temperature. If effects arising from the formation of cracks are ignored, permanent deformation in a material under indentation loading is thus seen as arising from a combination of instantaneous plasticity (which is not time-dependent) and creep (which is time-dependent). A material that undergoes elastic and non-time-dependent plastic deformation is called elastoplastic. A material that deforms elastically but exhibits time-dependent behavior is called viscoelastic. A material in which time-dependent plastic deformation occurs is viscoplastic[§§]. The term creep is often used to describe a delayed response to an applied stress or strain that may be a result of viscoelastic or viscoplastic deformation. In a nanoindentation test, the depth recorded at each load increment will be, in general, the addition of that due to the elastic–plastic properties of the material and that occurring to due creep, either viscoelastic or viscoplastic.

The physical mechanism of creep is of considerable interest. Li and Warren[9] propose that the highly concentrated indentation stress fields in the specimen material result in a chemical potential gradient that leads to a thermally activated diffusional flux of atoms flowing from the area beneath the indenter to the specimen surface and also along the interface between the indenter and specimen, even under what is nominally an elastic contact. The rate equation for indentation with a spherical indenter was found to be:

$$\frac{dh}{dt}=\frac{2(1+\nu)}{\pi}\frac{P}{\left[h(2R-h)\right]^{3/2}}\frac{D_v\Omega}{kT}\left[1+\frac{8\delta}{3\left[h(2R-h)\right]^{1/2}}\frac{D_i}{D_v}\right] \tag{7.3a}$$

§§ Strictly speaking, time-dependent plasticity is one where the deformation rate is fixed by the mechanical configuration of the specimen. It is a special case of the more general term of viscoplasticity.

and for a conical indenter:

$$\frac{dh}{dt} = \frac{2(1+\nu)}{\pi \tan^3 \alpha} \frac{P}{h^3} \frac{D_v \Omega}{kT} \left[1 + \frac{8\delta}{3h \tan \alpha} \frac{D_i}{D_v} \right] \tag{7.3b}$$

In Eqs. 7.3a and 7.3b, ν is the Poisson's ratio of the specimen, D_v is the diffusion coefficient within the specimen, D_i is the diffusion coefficient along the interface between the specimen and the indenter, Ω is the atomic volume of the specimen material, k is Boltzmann's constant, T is the absolute temperature, δ is the thickness of the interface along which diffusion occurs, α is the cone semi-angle, h is the depth of penetration beneath the specimen surface measured at the circle of contact (i.e., h in Eqs. 7.3a and 7.3b is the plastic depth h_p), and R is the indenter radius. This diffusion equation holds best when the indentation temperature is very much below the melting point of the specimen material. When the indentation temperature is more than one half of the specimen melting temperature, a simple steady-state power-law creep equation may be more appropriate.[10]

There are several detailed theoretical studies of viscoelastic indentation creep available in the literature.[11-16] In nanoindentation testing, it is of significant practical interest to measure the viscoelastic properties of materials and phenomenological approach based on spring and dashpot elements is usually sufficient to provide meaningful data for a wide range of materials.

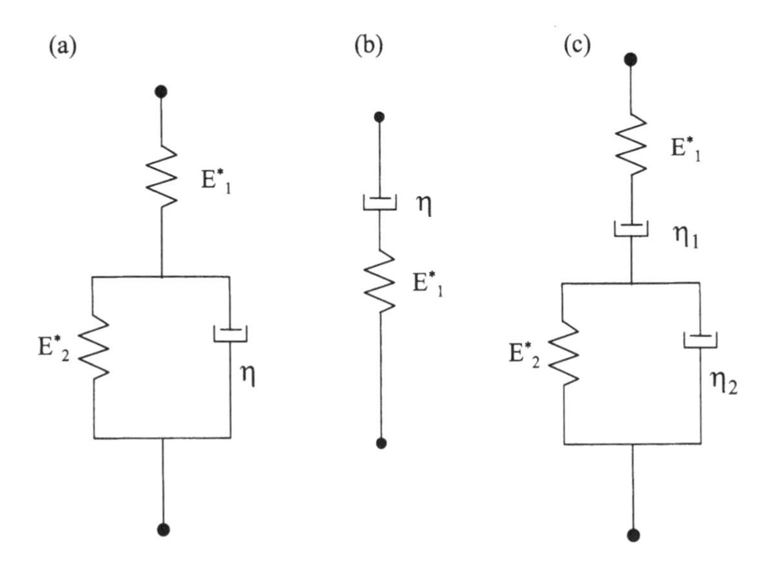

Fig. 7.7 (a) Three-element Voigt spring and dashpot representation of a viscoelastic solid (delayed elastic response) (b) Maxwell representation of a viscoelastic solid (steady creep) (c) Combined Maxwell-Voigt four-element model. .

A phenomenological approach focuses on the phenomena rather than the cause. Time dependent properties of materials are conventionally analyzed in terms of mechanical models such as the 3-element Voigt and 2-element Maxwell models shown in Fig. 7.7. Since most yield responses of materials are a result of shear, it is usual to express mechanical constants of the model elements in terms of the bulk modulus K and shear modulus G of the material that they represent. In these models, η is the viscosity term that quantifies the time-dependent property of the material. The elastic modulus is related to the shear modulus G and bulk modulus K by:

$$K = \frac{E}{3(1-2\nu)} \qquad G = \frac{E}{2(1+\nu)} \tag{7.3c}$$

The elastic response of models such as those shown in Fig. 7.7 is quantified by the storage modulus. The fluid-like response is quantified by the loss modulus. Since most yield responses of materials are a result of shear, it is usual to express mechanical constants of the model elements in terms of the bulk and shear moduli of the material that they represent, that is, the storage modulus is usually associated with the conventional shear modulus of the specimen. However, in an indentation test, the nature of the loading is a complex mixture of hydrostatic compression, tension, and shear. The storage modulus, as a result, is likely to be due to all three of these types of materials response. In this section, we shall, in the interests of simplicity, assume that the bulk of the elastic response is due to the conventional elastic modulus. The fluid-like response we shall call "viscosity" although in practice, viscosity is usually frequency and temperature dependent and not single-valued. Radok,[17] and Lee and Radok[18] have analyzed the viscoelastic contact problem using a correspondence principle whereby elastic constants in the elastic equations of contact are replaced with time dependent operators. For an incompressible material ($\nu = 0.5$), deformations are conveniently expressed in terms of the deviatoric stresses and strains and as such, the elastic modulus is replaced by time dependent forms of the quantity 2G.[19] In terms of E^*, for a spherical indenter for a three-element Voigt model (Fig. 7.7 (a)), for a steady applied load P_o, the depth of penetration increases with time according to:

$$h^{3/2}(t) = \frac{3}{4}\frac{P_o}{\sqrt{R}}\left[\frac{1}{E^*_1} + \frac{1}{E^*_2}\left(1 - e^{-t\frac{E^*_2}{\eta}}\right)\right] \tag{7.3d}$$

The square-bracketed term in Eq. 7.3d represents the time response of the mechanical model to a step load. A step increase in applied load results in an initial elastic displacement ($t = 0$) followed by a delayed increase in displacement to a maximum value at $t = \infty$ as shown in Fig. 7.8 (a).

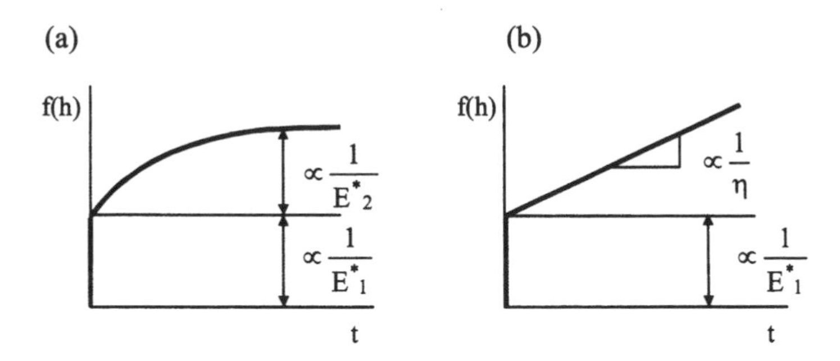

Fig. 7.8 Displacement response for a step increase in load for (a) Voigt model, (b) Maxwell model. $f(h) = h^2$ for cone, $f(h) = h^{3/2}$ for sphere.

A similar approach is appropriate for the case of a conical indenter in which we obtain, for the case of the three-element Voigt model:

$$h^2(t) = \frac{\pi}{2} P_o \cot\alpha \left[\frac{1}{E^*_1} + \frac{1}{E^*_2}\left(1 - e^{-t\frac{E^*_2}{\eta}}\right)\right] \tag{7.3e}$$

For the case of a Maxwell model Fig. 7.7 (b), the time-dependent depth of penetration for a spherical indenter is given by:

$$h^{3/2}(t) = \frac{3}{4}\frac{P_o}{\sqrt{R}}\left[\frac{1}{E^*_1} + \frac{1}{\eta} t\right] \tag{7.3f}$$

and for a conical indenter, we obtain:

$$h^2(t) = \frac{\pi}{2} P_o \cot\alpha \left[\frac{1}{E^*_1} + \frac{1}{\eta} t\right] \tag{7.3g}$$

It is fairly straight-forward to combine the above equations to provide expressions for the combined four-element model shown in Fig. 7.7 (c). In the case of a spherical indenter, we have:

$$h^{3/2}(t) = \frac{3}{4}\frac{P_o}{\sqrt{R}}\left[\frac{1}{E^*_1} + \frac{1}{E^*_2}\left(1 - e^{-t\frac{E^*_2}{\eta_2}}\right) + \frac{1}{\eta_1} t\right] \tag{7.3h}$$

and for a conical indenter:

$$h^2(t) = \frac{\pi}{2} P_o \cot\alpha \left[\frac{1}{E^*_1} + \frac{1}{E^*_2}\left(1 - e^{-t\frac{E^*_2}{\eta_2}}\right) + \frac{1}{\eta_1} t\right] \tag{7.3i}$$

It should be remembered that Eqs. 7.3d to 7.3i assume a step increase in load to P_o. Methods of superposition[20] can be used to analyze an arbitrary load history. The equations given above apply to loading whereby the contact radius increases with time. A more general approach for loading and unloading is given by Ting.[21] The equations given above are expressed in terms of the modulus E^* which is the combination of elastic modulus and Poisson's ratio of the specimen as is done in conventional nanoindentation analysis, and η is the viscosity term that quantifies the time-dependent property of the material. Eqs. 7.3d to 7.3i can be fitted to experimentally obtained creep data to provide values for E^* and η.

A popular motivation for modeling of this type is the behavior of materials at elevated temperatures.[12,22] Modeling of indentation creep for nanoindentation applications usually focuses on either constant load (creep[9]) or constant displacement (relaxation[23]) or both.[24,25] For example, Feng and Ngan[26] and Ngan and Tang[27] applied a Maxwell two-element model to the creep displacement at maximum load in a conventional load-displacement response and determined an equivalent expression for the contact stiffness that included the creep rate expressed as a displacement over time. Their work illustrates and quantifies the forward going "nose" that appears in the unloading curve in indentation experiments in which creep is significant. It is seen that for a hold period at maximum load followed by an unloading, the contact stiffness S, in the presence of creep, is found from:

$$\frac{1}{S} = \frac{1}{S_u} + \frac{\dot{h}_h}{|\dot{P}|} \tag{7.3j}$$

where S_u is the measured or apparent unloading stiffness (affected by creep), $\dot{h}_h$ is the creep rate at the end of the hold period and $\dot{P}$ is the initial unloading rate.

Cheng, Scriven and Gerberich[28] applied a method of functional equations to the viscoelastic contact problem in conjunction with a step increase in load, or displacement, to a three-element Voigt model to provide equations for steady creep, or relaxation, for the case of a rigid spherical indenter in contact with both incompressible and compressible materials. They found that the creep response at a step increase in load to a constant value for a spherical indenter for a Voigt model shown in Fig. 7.7 (a) could be expressed as:

$$h(t)^{3/2} = \frac{3P_o\left(1-\nu^2\right)}{4\sqrt{R}E_1}\left(A_c e^{-\alpha_c t} + B_c e^{-\beta_c t} + C_c\right) \tag{7.3k}$$

In Eq. 7.3k, P_o is the fixed load, R is the radius of the indenter and A_c, B_c, C_c, α_c and β_c are given by:

$$A_c = \frac{\left(1-\frac{t_c}{t_r}\right)\left(a-\frac{t_r}{t_c}\right)}{\left(b-\frac{t_r}{t_c}\right)}; B_c = \frac{(b-a)(b-1)}{b\left(b-\frac{t_r}{t_c}\right)}; C_c = \frac{a}{b}\frac{t_c}{t_r} \tag{7.3l}$$

$$\alpha_c = \frac{1}{t_c}; \beta_c = \frac{b}{t_r}$$

In Eq. 7.3l,

$$a = \frac{3K_1p_1 + 2q_op_1}{3K_1p_1 + 2q_1}; b = \frac{6K_1p_1 + q_op_1}{6K_1p_1 + q_1}; t_r = p_1; t_c = \frac{q_1}{q_o} \tag{7.3m}$$

where

$$p_1 = \frac{\eta}{G_1 + G_2}; q_o = \frac{2G_1G_2}{G_1 + G_2}; q_1 = \frac{2G_1\eta}{G_1 + G_2} \tag{7.3n}$$

In contrast, Oyen and Cook[29] presented a phenomenological approach which sought to include elasticity, viscosity, and plasticity in terms of modeling elements that represented the quadratic character of the contact equations rather than the intrinsic properties of the specimen material. Those workers found a very good agreement between the predicted and observed load-displacement curves for the bounding conditions of an elastic–plastic response (e.g. metals and ceramics) to viscoelastic deformation (e.g. elastomers).

Strojny and Gerberich[30] used a least squares method to fit the adjustable parameters to experimental results for polymeric thin films and found that the resulting elastic and shear moduli obtained for the films were comparable to those measured for the bulk materials. The method is applicable to both compressible and incompressible materials.

Syed Asif and Pethica[31] report nanoindentation creep tests on high purity Indium for the purpose of measuring the strain rate dependence on hardness, stress exponent and activation energy.

As mentioned previously, the depth recorded at each load increment in a nanoindentation test will be that arising from elastic–plastic, viscoelastic and viscoplastic deformations. In conventional nanoindentation testing, the instantaneous elastic and plastic deformations are usually considered. Elastic equations of contact are applied to the unloading data to find the depth of the circle of contact at full load (under elastic–plastic conditions). If there is a viscoelastic or viscoplastic response (i.e. creep), then this analysis is invalid. In extreme cases, the unloading curve has a negative slope which, if the standard unloading analysis is applied, results in a negative elastic modulus.[27,32]

When fitting creep curves to mechanical models, we must be clear about what properties are being measured. For time-dependent properties, we must be sure that we distinguish between viscoelasticity and viscoplasticity. If a step load is applied to an indenter in contact with a material and the resulting depth of penetration monitored, then a response similar to that shown in Fig. 7.8 (a) or (b) may be obtained. Fitting the appropriate equation (Eqs. 7.3d to 7.3i) to this data yields values for the storage and loss moduli. The accuracy of the results so obtained depends upon the rapidity with which the step increase in load is applied and the time-dependent nature of the specimen material. In practice, an increase in load is applied over a finite time period within which, for elastic–plastic materials, plastic deformation can occur quite rapidly and this causes the initial step response in displacement to be greater than that predicted, particularly when a sharp indenter is used. The resulting value of elastic modulus can be very much less than the nominal modulus of the material. In this case, the measured viscous component is likely to be that associated with viscoplasticity. For a step load with a spherical indenter on a material with a significant viscous response, the resulting analysis is likely to result in a reasonable measurement of the viscoelastic properties of the material.

While the use of simple mechanical models such as those shown in Fig. 7.7 allows some comparison to be made between specimens, the modeling of them is a phenomenological approach and subsequently offers very little in terms of a basic understanding of the physical mechanisms involved in the deformation. For example, time-dependent behavior often depends on the strain-hardening characteristics of the material which in turn may depend upon microstructural variables. Various constitutive laws[33] have been proposed that apply to many different types of materials and these should be investigated if a more detailed account of indentation creep is desired.

Some workers[32,34] have studied the effect of creep on the values of modulus and hardness obtained using conventional methods of nanoindentation analysis of the unloading response. The general conclusion is that the modulus so calculated is not reliable if the hold period at maximum load is too short due to bowing of the unloading response to larger depth values resulting from creep. As a general rule, Chudoba and Richter[32] found that the hold period at maximum load has to be long enough such that the creep rate has decayed to a value where the depth increase in one minute is less than 1% of the indentation depth. At this condition, the unloading slope is found to be free from the influence of creep and thus provides a measure of contact stiffness at the increased penetration depth. Much the same effect can be obtained by undertaking the indentation test very slowly.

References

1. A. Kulkarni and B. Bhushan, "Nano/picoindentation measurements on a single-crystal aluminium using modified atomic force microscopy," Mat. Lett. 29, 1996, pp. 221–227.
2. S. Akita, H. Nishijima, T. Kishida and Y. Nakayama, "Nanoindentation of polycarbonate using a carbon nanotube tip," Jpn. J. Appl. Phys. 39, 2000, pp. 7086–7089.
3. S.A. Syed Asif, R.J. Colton, and K.J. Wahl, "Nanoscale surface mechanical property measurements: Force modulation techniques applied to nanoindentation," in *Interfacial properties on the submicron scale*, J. Frommer and R. Overney, eds. ACS Books, Washington, DC, 2001, pp. 189–215.
4. S.A. Syed Asif, K.J. Wahl, and R.J. Colton, 'The influence of oxide and adsorbates on the nanomechanical response of silicon surfaces," J. Mater. Res. 15 2, 2000, pp. 546–553.
5. B.N. Lucas, W.C. Oliver, and J.E. Swindeman, "The dynamics of frequency specific, depth sensing indentation testing," Mat. Res. Soc. Symp. Proc. 522, 1998, pp. 3–14.
6. J.-L. Loubet, B.N. Lucas, and W.C. Oliver, in NIST Special Publication 896, *Conference Proceedings: International Workshop on Instrumented Indentation*, eds. D.T. Smith (NIST) 1995, pp. 31–34.
7. N.A. Burnham, S.P. Baker, and H.M. Pollock, "A model for mechanical properties of nanoprobes," J. Mater. Res. 15 9, 2000, pp. 2006–2014.
8. A.F. Bower, N.A. Fleck, A. Needleman, and N. Ogbonna, "Indentation of a power law creeping solid," Proc. Roy. Soc. A441, 1993, pp. 97–124.
9. W.B. Li and R. Warren, "A model for nano-indentation creep," Acta Metall. Mater. 41 10, 1993, pp. 3065–3069.
10. P.M. Sargent and M.F. Ashby, "Indentation creep," Mat. Sci. and Tech. 8, 1992, pp. 594–601.
11. R. Hill, B. Storåkers, A.B. Zdunek, "A theoretical study of the Brinell hardness test," Proc. Roy. Soc. A423, 1989, pp. 301–330.
12. T.R.G. Kutty, C. Ganguly and D.H. Sastry, "Development of creep curves from hot indentation hardness data," Scripta Materialia, 34 12, 1996, pp. 1833–1838.
13. R. Hill, "Similarity analysis of creep indentation tests," Proc. Roy. Soc. A436, 1992, pp. 617–630.
14. B. Storåkers and P.-L. Larsson, "On Brinell and Boussinesq indentation of creeping solids," J. Mech. Phys. Solids, 42 2, 1994 ,pp. 307–332.
15. M. Sakai, "Time-dependent viscoelastic relation between load and penetration for an axisummetric indenter," Phil. Mag. A 82 10, 2002, pp. 1841–1849.
16. S.Dj. Mesarovic and N.A. Fleck, "Spherical indentation of elastic-plastic solids," Proc. Roy. Soc. A455, 1999, pp. 2707–2728.
17. J.R.M. Radok, "Viscoelastic stress analysis," Q. Appl. Math. 15, 1957, pp. 198–202.
18. E.H. Lee and J.R.M. Radok, "The contact problem for viscoelastic solids," Trans. ASME Series E, J.App.Mech. 27, 1960, pp. 438–444.
19. K.L. Johnson, *Contact Mechanics*, Cambridge University Press, 1985.

20. L. Cheng, X. Xia, W. Yu, L.E. Scriven, and W.W. Gerberich, "Flat-punch indentation of a viscoelastic material," J. Polymer Sci. 38, 2000, pp. 10–22.
21. T.C.T. Ting, "The contact stresses between a rigid indenter and a viscoelastic half-space," Trans. ASME, J. App. Mech. 33, 1966, pp. 845–854.
22. M. Sakai and S. Shimizu, "Indentation rheometry for glass-forming materials," J. Non-Crystalline Solids, 282, 2001, pp. 236–247.
23. S. Shimizu, Y. Yanagimoto and M. Sakai, "Pyramidal indentation load-depth curve of viscoelastic materials," J. Mater. Res. 14 10, 1999, pp. 4075–4085.
24. X. Xia, A. Strojny, L.E. Scriven, W.W. Gerberich, A. Tsou, and C.C. Anderson, "Constitutive property evaluation of polymeric coatings using nanomechancial methods," Mat. Res. Symp. Proc. 522, 1998, 199–204.
25. K.B. Yoder, S. Ahuja, K.T. Dihn, D.A. Crowson, S.G. Corcoran, L. Cheng, and W.W. Gerberich, "Nanoindentation of viscoelastic materials: Mechanical properties of polymer coatings an aluminum sheets," Mat. Res. Symp. 522, 1998, pp. 205–210.
26. G. Feng and A.H.W. Ngan, "Effects of creep and thermal drift on modulus measurement using depth-sensing indentation," J. Mater. Res. 17 3, 2002, pp. 660 – 668.
27. A.H.W. Ngan and B. Tang, "Viscoelastic effects during unloading in depth-sensing indentation," J. Mater. Res. 17 10, 2002, pp. 2604–2610.
28. L. Cheng, L.E. Scriven, and W.W. Gerberich, "Viscoelastic analysis of micro- and nanoindentation," Mat. Res. Symp. Proc. 522, 1998, pp. 193–198.
29. M.L. Oyen and R.F. Cook, "Load-displacement behavior during sharp indentation of viscous-elastic-plastic materials," J. Mater. Res. 18 1, 2003, pp. 139–150.
30. A. Strojny and W.W. Gerberich, "Experimental analysis of viscoelastic behavior in nanoindentation," Mat. Res. Soc. Symp. Proc. 522, 1998, pp. 159–164.
31. S.A. Syed Asif and J.B. Pethica, "Nano-scale indentation creep testing at non-ambient temperatures," J.Adhesion, 67, 1998, pp. 153–165.
32. T. Chudoba and F. Richter, "Investigation of creep behavior under load during indentation experiments and its influence on hardness and modulus results," Surf. Coat. Tech. 148, 2001, pp. 191–198.
33. D. François, A. Pineau, and A. Zaoui, *Mechanical Behaviour of Materials*, Kluwer Academic Publishers, The Netherlands, 1998.
34. B.J. Briscoe, L. Fiori and E. Pelillo, "Nano-indentation of polymeric surfaces," J. Phys. D. Appl. Phys. 31, 1998, pp. 2395–2405.

Chapter 8
Nanoindentation of Thin Films

8.1 Introduction

One of the most popular applications of nanoindentation is the determination of the mechanical properties of thin films. In nanoindentation tests, the properties of the film may be measured without removing the film from the substrate as is done in other types of testing. The spatial distribution of properties, in both lateral and depth dimensions, may be measured, and a wide variety of films are amenable to the technique, from ion-implanted surfaces to optical coatings and polymer films. Apart from testing films in-situ, nanoindentation techniques can also be used for films made as free-standing microbeams or membranes.[1]

8.2 Testing of Thin Films

Both quantitative and qualitative information can be obtained from nanoindentation experiments on thin film systems. For example, comparison of load-displacement curves between coated and uncoated substrates often reveals changes in the elastic and plastic response of a system due to differences in surface treatment. The presence of discontinuities in the load-displacement response reveals information about cracking, delamination, and plasticity in the film and substrate. Some properties of thin films tested using nanoindentation techniques are given in Appendix 5.

The chief difficulty encountered in nanoindentation of thin films is to avoid unintentional probing of the properties of the substrate. To achieve this, it is common to restrict the maximum depth of penetration in a test to no more than 10% of the film thickness, although research suggests that this rule has no physical basis[2]. For indentations with a conical or pyramidal indenter, the indentation depth increases at the same rate as the radius of the circle of contact. Thus, for an indentation test on a thin film, the indentation scales with the ratio of the radius of the circle of contact divided by the film thickness a/t. The ratio of the penetration depth and the film thickness, h/t, can also be used as a scale parameter. The former is probably a more useful parameter for hard films on soft substrates since the contact radius is approximately equal to that of the hydrostatic core (see Section 1.4.2) beneath the indenter. For soft films, h/t is of more interest since it is a measure of how far the penetration depth has approached the substrate.

8.2.1 Elastic Modulus

For measurements of elastic modulus, influence from the substrate is unavoidable. Despite this, there are various treatments available to account for this.[3,4] King[5] evaluated the empirical treatment of Doerner and Nix[6] using the finite element method to arrive at an expression for the combined modulus of the film, substrate and indenter, E_{eff}:

$$\frac{1}{E_{eff}} = \frac{\left(1-\nu_f^2\right)}{E_f}\left(1-e^{-\alpha t/\sqrt{A}}\right) + \frac{\left(1-\nu_s^2\right)}{E_s}\left(e^{-\alpha t/\sqrt{A}}\right) + \frac{\left(1-\nu_i^2\right)}{E_i} \qquad (8.2.1a)$$

In Eq. 8.2.1a, the subscripts f, s, and i refer to the film, substrate, and indenter respectively. E_{eff} in Eq. 8.2.1a, in conjunction with the shape factor β (see Table 1.1) takes the place of E^* in Eq 3.2.4.1e. t is the film thickness, and the empirical constant α is required to be evaluated from a series of experimental results on films of known properties and thicknesses. King provides some representative values of α for different film thicknesses and indenter geometries.

Gao, Chiu, and Lee[7] used a moduli-perturbation method in which a closed-form solution results in an expression for the combined elastic modulus E_{eff} of the film/substrate combination:

$$E_{eff} = E_s + \left(E_f - E_s\right)I_o \qquad (8.2.1b)$$

where in Eq. 8.2.1b, I_o is a function of t/a (where t is the film thickness and a is the contact radius) given by:

$$I_o = \frac{2}{\pi}\tan^{-1}\frac{t}{a} + \left[\left(1-2\nu\right)\frac{t}{a}\ln\left(1+\frac{(t/a)^2}{(t/a)^2}\right) - \frac{t/a}{1+(t/a)^2}\right]\left(\frac{1}{2\pi(1-\nu)}\right) \qquad (8.2.1c)$$

In Eq. 8.2.1c, I_o is a weighting function that equals zero as the film thickness approaches zero, and approaches unity for large values of film thickness. Swain and Weppelmann used Eq. 8.2.1c to evaluate I_o for a range of elastic indentations on a TiN film on a silicon substrate, to obtain the combined film/substrate modulus E_{eff}, and bulk silicon for E_s. E_f could be estimated from:

$$\frac{E_{eff}}{E_s} = \left(1-I_o\right) + E_f I_o \qquad (8.2.1d)$$

These equations are of considerable importance since the measurement of modulus for the film is inherently connected with that of the substrate. As a very rough approximation, the stiffness of the two (or more) elastic elements comprising the thin film and substrate are essentially in series as shown in Fig. 8.1 (a). No matter how low of a force is applied to the indenter, there will always be a contribution from the substrate. However, in the indentation stress field, the indenter load is supported not only by direct compression in the vertical direction, as shown in Fig. 8.1 (a), but also by compressive stresses acting inwards from the sides. This makes the contribution from the film somewhat more than that expected from simple compression in the vertical direction only. Since there is always some elastic displacement of the substrate during an indentation test, the traditional 10% rule does not strictly apply for nanoindentation measurements for modulus determination of thin film systems.

Despite the fundamental difficulties in extracting the film modulus from the load-penetration depth data from first principles, it is a relatively straightforward procedure to undertake a series of tests with a conical or pyramidal indenter at differing depths and to plot the measured combined modulus E_{eff} against the scaling parameter a/t. The film modulus E_f is found by extrapolating the curve of best fit to these data to zero a/t as shown in Fig. 8.1 (b).

The analytical computation of the stress distribution of a thin film system in indentation loading has traditionally been very difficult and so this information is usually obtained by finite element methods. Schwarzer and co-workers[8,9] have developed an analytical procedure using a method of image charges. This treatment provides an elastic solution for the complete stress distribution in the film and substrate. Values of tensile or shear stresses that might subsequently lead to fracture or plastic deformation can be readily identified.

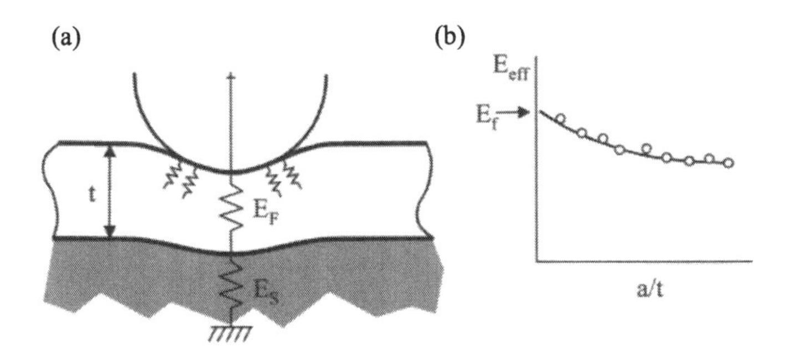

Fig. 8.1 (a) The contribution to E_{eff} from the film includes that from the localized support from the indentation stress field. (b) For a geometrically similar indenter, the indentation scales with a/t. The modulus of the film E_f is found by extrapolating measured values of E_{eff} to a/t = 0.

8.2.2 Hardness

The hardness value measured for a film and substrate combination is more difficult to quantify than the combined elastic modulus due to the complex nature of the plastic zone as it interacts with the substrate material. Research in this area is usually concerned with establishing a value for the critical indentation depth–that depth of penetration of the film, expressed as a percentage, below which the influence from the substrate is less than some desired amount.

Bückle[10] proposed that the composite hardness of a thin film system H_{eff} could be expressed as:

$$H_{eff} = H_s + \alpha\left(H_f - H_s\right) \tag{8.2.2a}$$

where H_s is the hardness of the substrate and H_f is the hardness of the film with α being an empirically derived parameter.

Jonsson and Hogmark[11] propose an area law of mixtures so that the hardness of the film can be extracted from the hardness of the film–substrate combination H_{eff} from:

$$H_{eff} = H_f \frac{A_f}{A} + H_s \frac{A_s}{A} \tag{8.2.2b}$$

where A_f and A_s are the relative parts of the contact carried by the film and the substrate, respectively, and A is the total contact area. A similar treatment, but based on a volume of deformation law of mixtures, was proposed by Burnett and Rickerby.[12,13] Tsui, Ross and Pharr[14] have proposed an empirical equation for correcting for pile-up effects for the case of soft materials on hard substrates which permits A in Eq. 8.2.2.b to be accurately determined without imaging.

For a soft film on a hard substrate, Bhattacharya and Nix[15] propose that the hardness of the film–substrate combination H_{eff} is determined by:

$$H_{eff} = H_s + \left(H_f - H_s\right)\exp\left(-\frac{Y_f}{Y_s}\frac{E_s}{E_f}\left(\frac{h}{t}\right)^2\right) \tag{8.2.2c}$$

where Y_f and Y_s are the material yield stresses for the film and substrate, respectively, and the exponential is a weighting function. For hard films on a soft substrate, the expression becomes:

$$H_{eff} = H_s + \left(H_f - H_s\right)\exp\left(-\frac{H_f}{H_s}\frac{Y_s}{Y_f}\sqrt{\frac{E_s}{E_f}}\frac{h}{t}\right) \tag{8.2.2d}$$

Stone, LaFontaine, Alexopolous, Wu, and Li[16] measured the hardness of aluminum films on silicon substrates as a function of depth of the indentation for different thicknesses of film and also altering the adhesion between the film and substrate by the prior deposition of a carbon layer. They found that the hardness increased as the indentation depth approached the interface.

More recently, Bull[17] proposed a hardness model based upon energy of deformation where the hardness is determined by the energy or work of indentation and the deformed volume of material. For a layered structure, if the bulk hardness H_o is known, then it is shown that the composite hardness is found by summing the value of H_o for each layer multiplied by its deformed volume, adding any surface energies of the interfaces, and then dividing by the total deformed volume.

Despite these analytical and sometimes empirical treatments, there is no one relationship as yet proposed which covers a wide range of materials behavior. In the absence of any rigorous relationship, the conventional 10% of the thickness rule appears to be that most generally used. Finite element modeling indicates that this can be relaxed to about 30% for the case of soft coatings on hard substrates.

8.2.3 Film Adhesion

The nanoindentation technique applied to thin film testing is not restricted to measuring mechanical properties of the film, but can also be applied to evaluate film adhesive strength. The strength of the bond between a film and substrate is intimately related to the residual stresses in the film and the stresses applied during service. Residual stresses are usually determined by a wafer bending technique using the Stoney[18] equation:

$$\sigma_f = \frac{E_s}{1-\nu_s}\frac{t_s^2}{6t_f}\left(\frac{1}{R}-\frac{1}{R_o}\right) \tag{8.2.3a}$$

where σ_f is the stress in the film, E_s and ν_s refer to the properties of the substrate, t_s is the substrate thickness, t_f is the film thickness, R_o is the initial radius of curvature, and R is the final radius of curvature of the wafer. Stresses in the GPa regime are commonly encountered for thin hard films. The adhesion is controlled by the strain energy release rate that has been calculated by Marshall and Evans[19]:

$$G = \frac{h\sigma_I^2\left(1-\nu^2\right)}{2E_f} + (1-\alpha)\frac{h\sigma_R^2(1-\nu)}{E_f} - (1-\alpha)\frac{h(\sigma_I-\sigma_R)^2(1-\nu)}{E_f} \tag{8.2.3b}$$

where σ_I is the indentation stress, σ_R is the residual stress, and α is a parameter between 0 and 1 depending on the degree of buckling in the film. If the film is not buckled, $\alpha = 1$ and the residual stresses do not contribute to G.

Alternatively, the strain energy release rate can be calculated using an annular plate analysis. This method is appropriate when the delaminated part of the film, under indentation loading, can be modeled as an annular plate where the inner radius is the contact radius a and the outer radius is the radius of the delaminated region c. If the internal radius is assumed to be under uniform pressure, the strain energy release rate is given as[20,21]:

$$G = \frac{2\left(1-\nu^2\right)\left(Y-H\right)^2 t}{E}\left(1+\nu+\left(1-\nu\right)\left(\frac{c}{a}\right)^2\right)^{-2} \qquad (8.2.3b)$$

where t is the film thickness and Y and H are the film yield stress and hardness respectively.

In a nanoindentation test, load may be applied to the indenter in a controlled manner so that the film is tested to failure. Features on the load displacement curve can be linked to mechanical failure processes[22,23] and one such example is indicated in Fig. 8.2. In Fig. 8.2, the difference in the areas under the curve for the uncracked and cracked specimen can be correlated to the strain energy released and thus used as a measure of fracture toughness if the length of the crack can be measured. In practice, the crack energy release rate is very small in comparison to the strain energy contained within the deformed zone and so the difference between the load-displacement curves for cracked and uncracked specimens is usually quite small unless there is significant delamination or spalling[24].

In some cases, indentation stresses alone may not be sufficient to cause delamination or failure of the film. Conventional tests for adhesion involve depositing an epoxy "superlayer" on the thin film system. The epoxy is cured at 180 °C and when cooled to room temperature, a tensile stress in the film is developed due to thermal expansion mismatch as shown in Fig. 8.3. The system is then further cooled until spontaneous delamination is observed. The test seeks a critical temperature at which delamination occurs as a measure of adhesion.

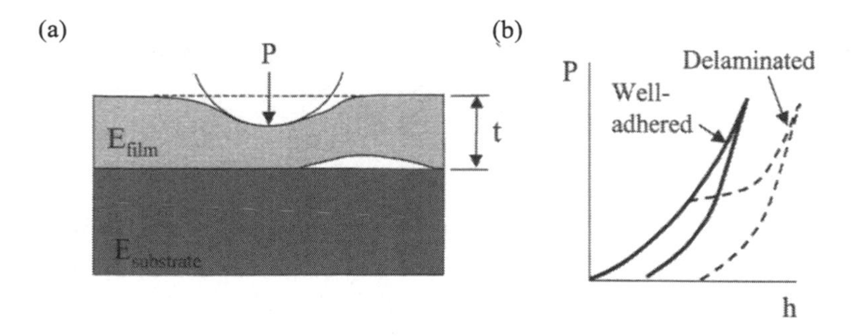

Fig. 8.2 Mechanical failure of a thin film system can lead to observable and identifiable features on a load-displacement curve in a nanoindentation test.

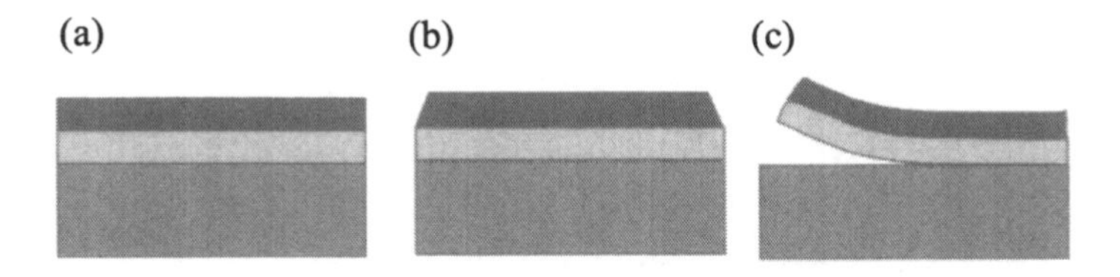

Fig. 8.3 Superlayer film testing technique. (a) An epoxy "superlayer" is deposited onto the thin film system. (b) The epoxy is cured at approximately 180 °C and then cooled until (c) a tensile stress in the film is developed due to thermal expansion mismatch and delamination occurs.

For tests with ductile or well-adhered films, stresses generated using the superlayer technique may not be sufficient to generate delamination. Kriese, Moody, and Gerberich[25] have used the superlayer technique in addition to a nanoindentation test to generate the required failure stresses in such cases. As reported by Volinsky, Moody, and Gerberich,[26] the technique has been extended to examining the effect of residual tensile and compressive stresses in thin film systems.

8.3 Scratch Testing

The scratch resistance of thin films and protective coatings is usually expressed in terms of their ability to withstand abrasion without fracturing. Scratch testing on a large scale enables films and coatings to be ranked according to the results of a particular test method. A typical scratch test involves a ramped load and the measure of performance is the critical load at which the surface fails. However, various modes of failure can be generated at different loads with different shapes of indenter.

Many nanoindentation instruments can be configured to operate in a scratch testing mode. As shown in Fig. 8.4, in this mode of operation, a normal force F_N is applied to the indenter, while at the same time, the specimen is moved sideways. In some instruments. An optional force transducer can be used to measure the friction, or tangential force F_T. In some cases, a lateral force F_L, normal to F_T can also be applied.[27]

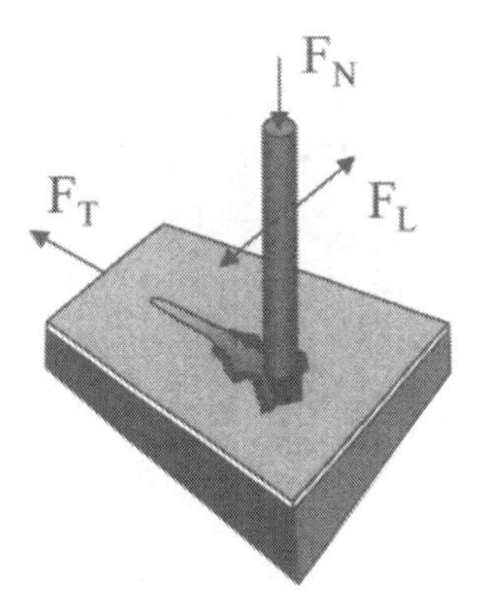

Fig. 8.4 Configuration of an indentation scratch test. The normal force F_N can be held at a constant value or ramped up or down while the specimen is moved sideways by a tangential force F_T. In addition, a lateral force F_L can sometimes also be applied.

Scratch tests on a micron scale were initially performed inside the chamber of an SEM.[28] The stylus or indenter in these tests was typically electro-polished tungsten tips with a radius of about 1 μm. The "scratch hardness" was defined as the track width of the scratch divided by the diameter of the scratch tip.

The physical meaning of the results of such a scratch test are fairly open to interpretation. In modern methods, the applied normal force is ramped up in value while the specimen is moved in a sideways direction and the minimum force F_C at which failure occurs is an indication of scratch resistance. The detection of this critical load can be determined using a variety of techniques such as optical microscopy, acoustic emission, and an analysis of the coefficient of friction, the latter method requiring a measurement of the tangential force F_T as well as the applied normal force F_N. The coefficient of friction μ can be readily calculated from:

$$\mu = \frac{F_T}{F_N} \tag{8.3a}$$

A diamond sphero-conical indenter with a tip radius of 200 μm is usually used as the stylus. The use of the critical load as a measure of scratch resistance is complicated by its dependence on scratching speed, loading rate, tip radius, substrate hardness, film thickness, film and substrate roughness, friction coefficient, and friction force.[29] Despite these difficulties, the method allows comparative tests to be performed with some degree of confidence to mechanical performance. Figures 8.5 and 8.6 show the results of a scratch test on multilayer Al/TiN/SiO film on a silicon substrate for a constant applied normal load of 30 mN using a 20 μm scratch tip. The circled area in both figures is an indication of an area of poor adhesion of the film.

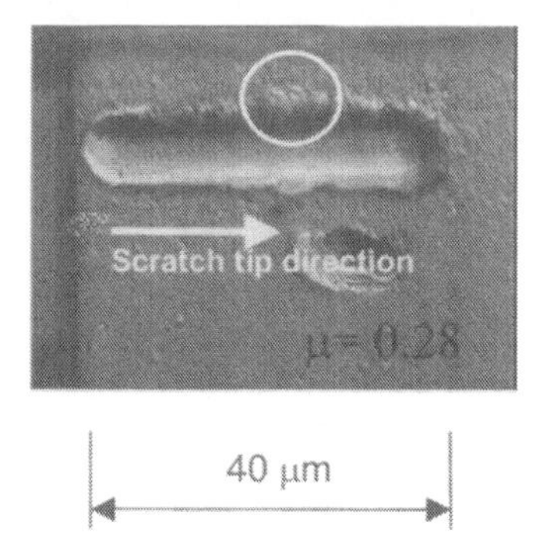

Fig. 8.5 Scratch test on a multilayer Al/TiN/SiO film on a silicon substrates for three different processing conditions for an applied normal load of 30 mN using a 20 μm scratch tip. The tip moved in a left to right direction as shown in the figure. The coefficient of friction is shown. The circled area indicates an area of poor adhesion of the film (Courtesy CSIRO).

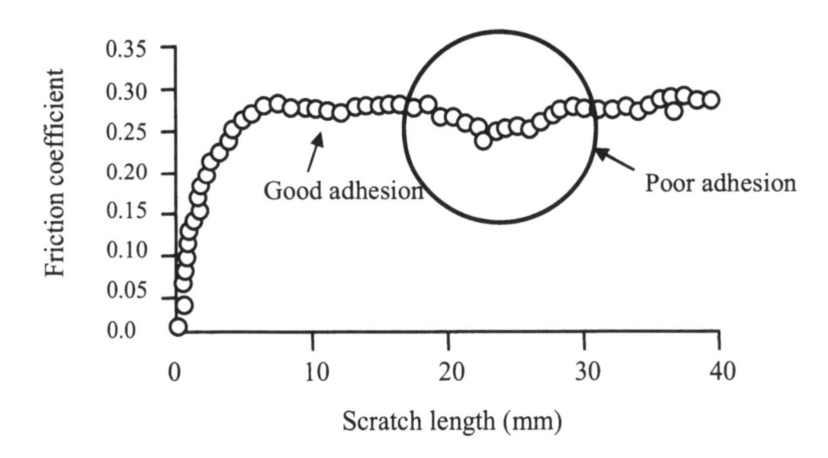

Fig. 8.6 Plot of coefficient of friction vs scratch length for specimen shown in Fig. 8.5. A marked decrease in the coefficient of friction is an indication of poor adhesion as shown in the circled area of Fig. 8.5 (Courtesy CSIRO).

In a ramped load scratch test, there is usually observed a transition from elastic to plastic deformation in the surface, and if the surface is a thin film, delamination eventually occurs. For scratches made on very soft materials, the shape of the indenter often affects the visibility of the scratch owing to an increase in piling-up along the length of the scratch. Jardret and Oliver[30] demonstrate the differences in piling-up with a Berkovich indenter with the differences in piling-up with a Berkovich indenter with the blunt face of the indenter facing forward, and the edge facing forward, on an automotive paint film.

Similar results obtained by Enders, Grau, and Berg[31] with sol-gel films on fused silica were interpreted in terms of an analytical model incorporating both sliding and ploughing friction terms.[32] Interestingly, this latter work treated the contact at very low loads as a single asperity contact in which surface adhesive forces (see Appendix 2) were included. The sliding part of the friction coefficient is given as:

$$\mu_S = \pi\left(\frac{3}{4E^*}\right)^{2/3} TR^{2/3}F_N^{-1/3}\left[1+\frac{F_A}{F_N}+\sqrt{\left(1+\frac{F_A}{F_N}\right)^2-1}\right]^{2/3} \quad (8.3b)$$

where T is the interfacial shear strength found from the tangential force F_T divided by the contact area, R is the radius of the indenter, F_A and F_N are the adhesive and normal forces respectively. The ploughing part as:

$$\mu_p = KP^*F_N^{(2-n)/n} \quad (8.3c)$$

where n and K are constants that depend upon the shape of the indenter. The total friction coefficient is the sum of these:

$$\mu = \mu_S + \mu_p \quad (8.3d)$$

Both the sliding and ploughing parts of the friction are functions of the normal force F_N. The sliding part of the friction coefficient dominates the process at low values of F_N and decreases with increasing load. The ploughing part of the friction is shown to dominate the process where, during the transition, the total friction coefficient passes through a minimum at a critical value of normal force F_N.

Scratch testing is of course closely related to the field of tribology. Tribological testing usually involves techniques such as pin-on-disk, ball-on-disk, and pins, rings, or disks on disks leading to measurements of friction force and friction coefficient, adhesion force of films, wear rate, contact resistance and acoustic emission of fracture events. While conventional nanoindentation instruments do not generally offer these capabilities, there is considerable overlap with some tribology instruments offering a nanoindentation mode of operation.

References

1. T.P. Weihs, S. Hong, J.C. Bravman, and W.D. Nix, "Mechanical deflection of cantilever microbeams: A new technique for testing the mechanical properties of thin films," J. Mater. Res. 3 5, 1988, 931–942.
2. J.L. Hay, M.E. O'Hern, and W.C. Oliver, "The importance of contact radius for substrate-independent property measurement of thin films," Mat. Res. Soc. Symp. Proc. 522, 1998, pp. 27–32.
3. M.G.D. El-Sherbiney and J. Halling, "The Herztian contact of surfaces covered with metallic films," Wear, 40, 1976, pp. 325–337.
4. J.A. Ogilvy, "A parametric elastic model for indentation testing of thin films," J. Phys. D: Appl. Phys. 26, 1993, pp. 2123–2131.
5. R.B. King, "Elastic analysis of some punch problems for a layered medium," Int. J. Solids Structures, 23 12, 1987, pp. 1657–1664.
6. M.F. Doerner and W.D. Nix, "A method of interpreting the data from depth-sensing indentation instruments," J. Mater. Res. 1 4, 1986, pp. 601–609.
7. H. Gao, C-H Chiu, and J. Lee, "Elastic contact versus indentation modeling of multi-layered materials," Int. J. Solids Structures, 29 20, 1992, pp. 2471–2492.
8. N. Schwarzer, M. Whittling, M. Swain, and F. Richter, "The analytical solution of the contact problem of spherical indenters on layered materials: Application for the investigation of TiN films on silicon," Thin Solid Films, 270 1-2, 1995, pp. 371–375.
9. N. Schwarzer, "Coating design due to analytical modelling of mechanical contact problems on multilayer systems," Surf. Coat. Technol. 133, 2000, pp. 397–402.
10. H. Bückle, in J.W. Westbrook and H. Conrad, eds. *The Science of Hardness Testing and its Applications*, American Society for Metals, Metals Park, OH, 1973, pp. 453–491.
11. B. Jonsson and S. Hogmark, "Hardness measurements of thin films," Thin Solid Films, 114, 1984, pp. 257–269.
12. P.J. Burnett and D.S. Rickerby, "The mechanical properties of wear-resistance coatings I: Modelling of hardness behaviour," Thin Solid Films, 148, 1987, pp. 41–50.
13. P.J. Burnett and D.S. Rickerby, "The mechanical properties of wear-resistance coatings II: Experimental studies and interpretation of hardness," Thin Solid Films, 148, 1987, pp. 51–65.
14. T.Y. Tsui, C.A. Ross, and G.M. Pharr, "A method for making substrate-independent hardness measurements of soft metallic films on hard substrates by nanoindentation," J. Mater. Res. 18 6, 2003, pp. 1383–1391.
15. A.K. Bhattacharya and W.D. Nix, "Finite element simulation of indentation experiments," Int. J. Solids Structures, 24 12, 1988, pp. 1287–1298.
16. D. Stone, W.R. LaFontaine, P. Alexopolous, T.-W. Wu, and Che-Yu Li, "An investigation of hardness and adhesion of sputter-deposited aluminium on silicon by utilizing a continuous indentation test," J. Mater. Res. 3 1, 1988, pp. 141–147.
17. S.J. Bull, "Modelling of the mechanical and tribological properties of coatings and surface treatments," Mat. Res. Symp. Proc. 750, 2003, pp. Y6.1.1–Y6.1.12.

18. G.G. Stoney, "The tension of metallic films deposited by electrolysis," Proc. Roy. Soc. A9, 1909, pp. 172–175.
19. D.B. Marshall and A.G. Evans, "Measurement of adherence of residually stressed thin films by indentation mechanics of interface delamination," J. Appl. Phys. 56 10, 1984, pp. 2632–2638.
20. L.G. Rosenfeld, J.E. Ritter, T.J. Lardner, and M.R. Lin, "Use of the microindentation technique for determining interfacial fracture energy," J. Appl. Phys. 67 1990, pp. 3291–3296.
21. M.D. Thouless, Acta Metall. 36, 1988, pp. 3131
22. M.V. Swain and J. Mencik, "Mechanical property characterization of thin films using spherical tipped indenters," Thin Solid Films, 253, 1994, pp. 204–211.
23. A.J. Whitehead and T.F. Page, "Nanoindentation studies of thin film coated systems," Thin Solid Films, 220, 1992, pp. 277–283.
24. M.D. Thouless, "An analysis of spalling in the microscratch test," Eng. Fract. Mech. 61, 1998, pp. 75–81.
25. M.D. Kriese, N.R. Moody, and W.W. Gerberich, "Effects of annealing and interlayers on the adhesion energy of copper thin films to SiO_2/Si substrates," Acta Mater. 46, 1998, pp. 6623–6630.
26. A. A. Volinsky, N.R. Moody, and W.W. Gerberich, "Superlayer residual stress effect on the indentation adhesion measurements," Mat. Res. Soc. Symp. Proc. 594, 2000, pp. 383–388.
27. J. Sekler, P.A. Steinmann, and H.E. Hintermann, "The scratch test: Different critical load determination techniques," Surface and Coatings Technology, 36, 1988, pp. 519–529.
28. N. Gane and J. Skinner, "The friction and scratch deformation of metals on a micro scale," Wear, 24, 1973, pp. 207–217.
29. P.A. Steinmann, Y. Tardy, and H.E. Hintermann, "Adhesion testing by the scratch test method: The influence of intrinsic and extrinsic parameters on the critical load," Thin Solid Films, 154, 1987, pp. 333–349.
30. V.D. Jardret and W.C. Oliver, "Viscoelastic behaviour of polymer films during scratch test: A quantitative analysis," Mat. Res. Soc. Symp. Proc. 594, 2000, pp. 251–256.
31. S. Enders, P. Grau, and G. Berg, "Mechanical characterization of surfaces by nanotribological measurements of sliding and abrasive terms," Mat. Res. Soc. Symp. Proc. 594, 2000, pp. 531–536.
32. F.P. Bowden and D. Tabor, *The Friction and Lubrication of Solids*, Oxford University Press, Oxford, 1950.

Chapter 9
Other Techniques in Nanoindentation

9.1 Introduction

Nanoindentation has proven to be a very versatile method of mechanical testing. It is often considered to be non-destructive in the sense that the indentations are in general, too small to be visible to the naked eye and, for the most part, the test does not impair the structural integrity of the specimen. Compared to the previous chapters, we now turn to a discussion of various unusual and advanced methods of testing that illustrate the versatility of the method.

9.2 Acoustic Emission Testing

When nanoindentation techniques are used to investigate cracking and delamination of thin films, acoustic emission (AE) can sometime be used with some advantage. The object of the acoustic emission measurement is to record the amplitude and time at which specific events occur during the application of load to the indenter. Examples of such events are substrate cracking, delamination, film cracking, phase transformations, and slippage beneath the indenter.

The acoustic emission sensor is typically a high-resolution, low-noise device with a high resonant frequency (≈200 – 300 kHz). This makes the sensor rather insensitive to mechanical vibrations from the environment. The sensor detects the elastic waves generated by the release of stored strain energy within the loaded system as some failure event occurs and is usually mounted either on the specimen or on the indenter shaft. The signal is amplified by a very high input impedance amplifier with a gain of approximately 100 db. The raw signal is typically filtered before presentation as either an rms voltage or an accumulation of "counts."

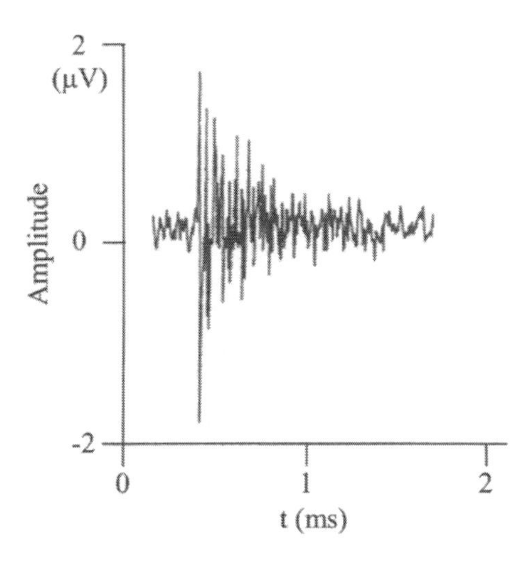

Fig. 9.1 Acoustic emission signal burst from a nanoindentation test (Courtesy CSIRO).

The acoustic emission signal typically contains a burst-type signal as shown in Fig. 9.1. Each signal burst corresponds to a damage event. The amplitude of the burst signal is an indication of the magnitude of the damage event. Shiwa, Weppelmann, Munz, Swain, and Kishi[1] compared features on a load-displacement curve obtained with a spherical indenter on silicon with recorded acoustic emission signals during an indentation test and identified five distinct stages of behaviour during the complete cycle from loading to final unloading. Events such as pressure-induced phase transformations and cracking were identified.

A quantitative approach to this type of testing was undertaken by Sekler, Steimann, and Hintermann,[27] who integrated the acoustic emission signal obtained on scratch tests on an TiC-coated cemented carbide specimen to produce an energy spectrum of events as a function of applied load. The critical load was determined by drawing a line through the average slope of the increasing energy data and reading off the intercept with the horizontal force axis. These authors found that despite the possibility of obtaining an objective measure of the critical load in a scratch test, it is frequently necessary for the output of the acoustic emission sensor to be adjusted by the operator so as to coincide with the damage event of interest before the technique can be used in an automated fashion.

In the work described above, the acoustic emission sensor was typically mounted on the side of the specimen mount. The disadvantage in this arrangement is that the acoustic emission signal is affected by the specimen dimensions due to multiple reflections. Further, the sensors used were those typically used in large scale deformations and so are limited in their ability to detect very small-scale phenomena such as would be of interest in nanoindentation testing. Tymiak, Daugela, Wyrobek and Warren[2] undertook a significant development of the technique through the use of a very small-scale acoustic emission sensor mounted directly on the indenter. Using this apparatus, these authors were able to demonstrate waveforms independent of the specimen and indenter size, a linear relationship between released elastic energies and detection of release in elastic energy of 2.5×10^{-14} J, some two orders of magnitude lower than previously reported. The work is of particular importance since it demonstrates, with some restrictions, that acoustic emission waveforms can be used to distinguish between different indentation-induced processes such a slip and twinning in different orientations of single-crystal sapphire.

9.3 Constant Strain Rate Testing

In earlier chapters, it was mentioned that the representative strain for testing with a Berkovich, Vickers, cube corner, or conical indenter was a constant owing to geometrical similarity of the contact. For a spherical indenter, Tabor[3] showed that the quantity 0.2a/R is a measure of the representative strain and increases as the depth of penetration increases. This empirical result, proposed in 1951, was the subject of a rigorous theoretical treatment by Hill, Storåkers, and Zdunek[4] in 1989 and an experimental study by Chaudhri[5] in 1993. These treatments are concerned with solids that show no dependence on the rate of application of strain. In some non-linear solids, such as those that follow a power-law creep response, the mechanical response of the material depends upon the rate of application of strain:

$$\sigma = K\dot{\varepsilon}^{m} \tag{9.3a}$$

where K and m are material constants. Several attempts have been made to extend the results of rate-independent materials to power-law creeping solids, the most rigorous perhaps being that of Bower, Fleck, Needleman and Ogbonna[6] and also Storåkers and Larsson.[7] The most popular treatment is that given by Mayo and Nix,[8] who define the strain rate for such tests as the indenter displacement velocity divided by the plastic depth:

$$\dot{\varepsilon} = \frac{dh}{dt}\frac{1}{h} \tag{9.3b}$$

where h in Eq. 9.3b is, strictly speaking, h_p. However, in soft metals, the elastic displacement h_e is very small compared to h_p and so the plastic depth can be approximated by the total depth of penetration. Equation 9.3b shows that the stress, or the mean contact pressure, decreases as the depth of penetration increases according to the power law in Eq. 9.3a. Precautions should be taken in tests on strain-hardening materials to minimize the depth dependence on hardness arising from geometrically necessary dislocations (see Section 4.7). For this reason, comparative testing of material to determine their strain rate sensitivities should be carried out at a fixed chosen depth of penetration. The strain rate sensitivity m is defined as[9]:

$$m = \frac{\partial \ln H}{\partial \ln \dot{\varepsilon}} \tag{9.3c}$$

and is a measure of the sensitivity of the hardness of a material to changes in strain rate. Cheng and Cheng[10] show that the hardness measured in an indentation test approaches a constant value when the strain rate is held constant and this can be obtained when the load rate divided by the load is held constant:

$$\frac{\dot{h}}{h} \propto \frac{\dot{P}}{P} \tag{9.3d}$$

In nanoindentation testing, it is sometimes desirable to undertake measurements on power-law creeping solids such that the conditions of Eq. 9.3d are satisfied during the application of load (and the unloading) so that meaningful value for hardness, independent of rate effects within the specimen, to be obtained. Typical tests in which strain rate is important are on specimens with a significant viscous component, but the issue does have some importance for some solids in which the deformation is of an unusual nature. For example, Schuh, Nieh and Kawamura[11] show that load-displacement curves on a bulk metallic glass exhibit a series of pop-in events corresponding to the motion of individual shear bands within the material only at low values of strain rate. Chinh, Horváth, Kovács, Lendvai[12] have studied plastic instabilities in metal alloys and bulk metallic glasses that result from a negative strain rate sensitivity originating mainly from the interaction between mobile dislocations and diffusing solute atoms during plastic deformation, a phenomenon called dynamic strain ageing. This type of deformation manifests itself as periodic serrations or steps in the load-displacement curve. The occurrence and development of the steps depends upon the loading rate, the composition of the alloy, and in some materials, the orientation of the grains with respect to the direction of indentation.

The issue of strain rate and its effect on the results of an indentation test are inextricably connected with the phenomenon of creep (see Chapter 7).

9.4 Fracture Toughness

Nanoindentation can be used to evaluate the fracture toughness of materials and interfaces in a similar manner to that conventionally used in larger scale testing. During loading, tensile stresses are induced in the specimen material as the radius of the plastic zone increases. Upon unloading, additional stresses arise as the elastically strained material outside the plastic zone attempts to resume its original shape but is prevented from doing so by the permanent deformation associated with the plastic zone. There exists a large body of literature on the subject of indentation cracking with Vickers and other sharp indenters. In this section, the method by which fracture toughness is evaluated from measurements of the sizes of surface cracks is reviewed.

Generally, there are three types of crack, and they are illustrated in Fig. 9.2. Radial cracks are "vertical" half-penny type cracks that occur on the surface of the specimen outside the plastic zone and at the corners of the residual impression at the indentation site. These radial cracks are formed by a hoop stress and extend downward into the specimen but are usually quite shallow.

Lateral cracks are "horizontal" cracks that occur beneath the surface and are symmetric with the load axis. They are produced by a tensile stress and often extend to the surface, resulting in a surface ring that may lead to chipping of the surface of the specimen. Median cracks are "vertical" circular penny cracks that form beneath the surface along the axis of symmetry and have a direction aligned with the corners of the residual impression. Depending on the loading conditions, median cracks may extend upward and join with surface radial cracks, thus forming two half-penny cracks that intersect the surface as shown in Fig. 9.2 (d). They arise due to the action of an outward stress. The exact sequence of initiation of these three types of cracks is sensitive to experimental conditions. However, it is generally observed that in soda-lime glass loaded with a Vickers indenter, median cracks initiate first. When the load is removed, the elastically strained material surrounding the median cracks cannot resume its former shape owing to the presence of the permanently deformed plastic material and this leads to a residual impression in the surface of the specimen.

Residual tensile stresses in the normal direction then produce a "horizontal" lateral crack that may or may not curve upward and intersect the specimen surface. Upon reloading, the lateral cracks close and the median cracks reopen. For low values of indenter load, radial cracks also form during unloading (in other materials, radial cracks may form during loading). For larger loads, upon unloading, the median cracks extend outward and upward and may join with the radial cracks to form a system of half-penny cracks, which are then referred to as "median/radial" cracks. In glass, the observed cracks at the corners of the residual impression on the specimen surface are usually fully formed median/radial cracks.

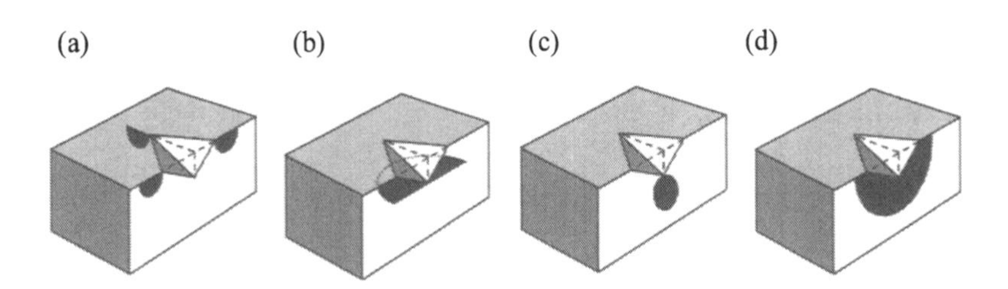

Fig. 9.2 Crack systems for Vickers indenter: (a) radial cracks, (b) lateral cracks, (c) median cracks, (d) half-penny cracks (after reference 13).

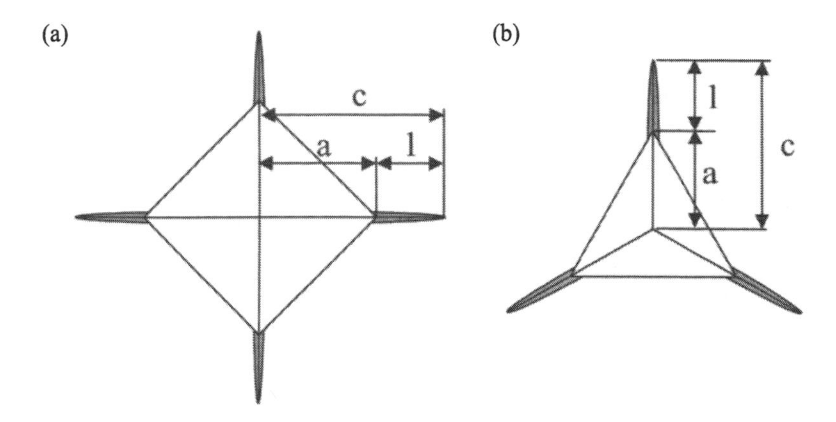

Fig. 9.3 Crack parameters for Vickers and Berkovich indenters. Crack length c is measured from the center of contact to end of crack at the specimen surface (After reference 13).

It is the radial and lateral cracks that are of particular importance, since their proximity to the surface has a significant influence on the fracture strength of the specimen. Fracture mechanics treatments of these types of cracks seek to provide a measure of fracture toughness based on the length of the radial surface cracks. Attention is usually given to the length of the radial cracks as measured from the corner of the indentation and then radially outward along the specimen surface as shown in Fig. 9.3.

Palmqvist[14] noted that the crack length "l" varied as a linear function of the indentation load. Lawn, Evans, and Marshall[15] formulated a different relationship, where they treated the fully formed median/radial crack and found the ratio $P/c^{3/2}$ (where c is measured from the centre of contact to the end of the corner radial crack) is a constant, the value of which depends on the specimen material. Fracture toughness is found from:

$$K_c = k\left(\frac{E}{H}\right)^n \frac{P}{c^{3/2}} \tag{9.4a}$$

where k is an empirical calibration constant equal to 0.016 and n = ½.

Various other studies have since been performed, and Anstis, Chantikul, Lawn, and Marshall[16] determined n = 3/2 and k = 0.0098. In 1987, Laugier[17] undertook an extensive review of previously reported experimental results and determined that:

$$K_c = x_v (a/l)^{1/2}\left(\frac{E}{H}\right)^{2/3} \frac{P}{c^{3/2}} \tag{9.4b}$$

where a is measured from the centre of the contact to the corner of the impression and l is measured from the corner of the impression to the end of the crack. With $x_v = 0.015$, Laugier showed that the radial and half-penny models make almost identical predictions of the dependence of crack length on load. Experiments show that the term $(a/l)^{1/2}$ shows little variation between glasses (median/radial) and ceramics (radial).

Although the vast majority of toughness determinations using indentation techniques are performed with a Vickers diamond pyramid indenter, the Berkovich indenter has particular usefulness in nanoindentation work. However, the loss of symmetry presents some problems in determining specimen toughness because half-penny cracks can no longer join two corners of the indentation. Ouchterlony[18] investigated the nature of the radial cracking emanating from a centrally loaded expansion star crack and determined a modification factor for stress intensity factor to account for the number of radial cracks formed:

$$k_1 = \sqrt{\frac{n/2}{1 + \frac{n}{2\pi}\sin\frac{2\pi}{n}}} \tag{9.4c}$$

As proposed by Dukino and Swain,[19] this modification has relevance to the crack pattern observed from indentations with a Berkovich indenter. The ratio of k_1 values for $n = 4$ (Vickers) and $n = 3$ (Berkovich) is 1.073 and thus the length of a radial crack (as measured from the center of the indentation to the crack tip) from a Berkovich indenter should be $1.073^{2/3} = 1.05$ that from a Vickers indenter for the same value of K_1. The Laugier expression can thus be written:

$$K_c = 1.073 x_v (a/l)^{1/2} \left(\frac{E}{H}\right)^{2/3} \frac{P}{c^{3/2}}. \tag{9.4d}$$

For very small penetrations in small volumes of specimen, cracks sizes can be difficult to measure directly, even if a cube-corner indenter is used to assist in the development of relatively large cracks at low loads. Field, Swain and Dukino[20] proposed a method of measurement of fracture toughness under these conditions by an instrumented approach. In this method, it was recognised that cracking is accompanied by an increase in penetration depth as shown in the load-displacement response. This "pop-in" event is thought to signal the nucleation of a median crack at the boundary of the plastic zone below the point of contact with the indenter. These authors determined the difference between the maximum actual penetration depth (with pop-in) and the anticipated penetration depth (with no pop-in and as determined from $P \propto h^2$ relationship) to calculate a value of crack length for use in Eq. 9.4d.

9.5 High-Temperature Nanoindentation Testing

An understanding of the wear behaviour for many structural components can best be obtained from tests that are performed at a temperature corresponding to the in-service temperature of the specimen. To this end, there have been a number of studies reported that are concerned with the development and use of hot hardness testers.[21–25] Experiments[25] show that, in general, the hardness of materials generally decreases with increasing temperature, especially at temperatures above 1000 °C. Hot hardness testing can also be used to evaluate creep properties of materials.[26]

High-temperature indentation on the nanometre scale is a less developed procedure than macroscopic hot hardness testing although the technique is now routinely commercially available. In one apparatus[27], nanoindentation was performed in a SEM under vacuum conditions. Syed Asif and Pethica[28] have combined a Peltier device with a nanoindentation instrument to measure creep and strain-rate dependence of indium at a maximum temperature of 60 °C. This device has the capability of testing at temperatures below room temperature to a minimum of approximately −5 °C.

Indentation testing at high temperatures brings with it several practical difficulties. Not only is the indenter material itself of some concern, but also the method of securely mounting it so as to withstand the indentation load. Figure 9.4 shows a Berkovich diamond indenter mounted by mechanical means only in a molybdenum chuck. This indenter is rated to approximately 750 °C in air.

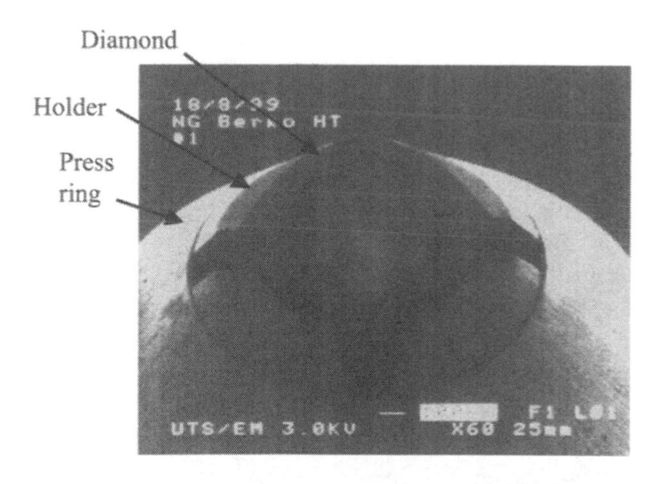

Fig. 9.4 Berkovich diamond pyramid indenter mounted in a molybdenum chuck designed for high-temperature indentation testing. The diamond "log" is held in position by the "press ring" (Courtesy CSIRO).

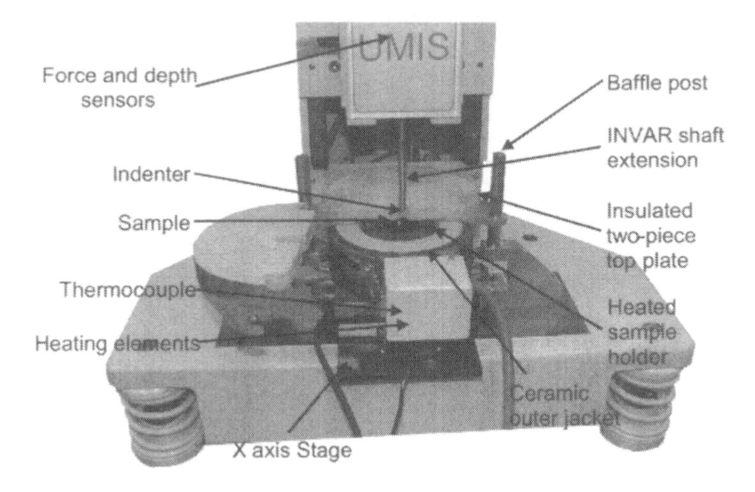

Fig. 9.5 Hot stage accessory fitted to a nanoindentation test instrument. A water-cooled jacket surrounds an insulated sample chamber that is heated using a temperature-controlled heating elements (Courtesy CSIRO).

Some nanoindentation instruments[29] offer a hot stage accessory that raises the temperature of the specimen to a high temperature (100 – 500 °C) where the convenience of ambient pressure testing is somewhat offset by the associated thermal insulation requirements to minimise thermal drift. Precise temperature control is required using a servo-type of temperature controller that can hold the temperature of the specimen to less 0.1 °C of the set point with a relatively long time cycle. The inevitable thermal drift that arises when the indenter is bought into contact with the specimen is a particular problem in high temperature nanoindentation testing. In some cases a separate tip heater is used to bring the indenter to the same temperature as the specimen prior to contact, thus minimising changes in dimensions of the tip and specimen when contact is made. In another apparatus,[30] the tip is bought into contact with the surface and held at the initial contact force for enough time to allow thermal equilibrium. Such a procedure requires a force feedback control loop that ensures that the load is held constant during the equilibrium process that can last 30 minutes or more. The small dimensions of the contact in nanoindentation experiments may lead to a thermal contact resistance that results in a steady-state temperature difference between the indenter and specimen. This in turn leads to changes in the dimensions of the contacting surfaces (thermoelastic contact). Experiments indicate that the effect does not affect the results to an appreciable extent, perhaps being due to the high thermal conductivity of the diamond indenter. Once thermal equilibrium is established, force and displacement data is collected during an indentation test in the normal manner. It is beneficial to include a hold period in the test cycle in order to evaluate thermal drift or creep during the load and unload sequence.

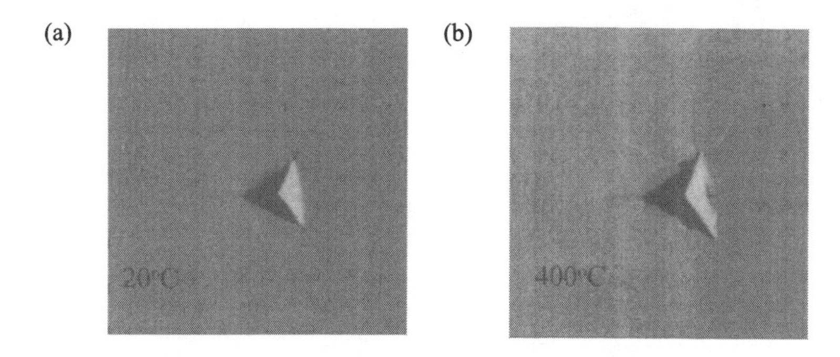

Fig. 9.6 Residual impressions made in sapphire at (a) 20 °C and (b) 400 °C with a Berkovich indenter at a maximum load of 1 N. Note the presence of cracks in the specimen indented at high temperature (Courtesy CSIRO).

Figure 9.6 show AFM images of residual impressions made in sapphire at 20 °C and 400 °C with a Berkovich indenter at a maximum load of 1 N. Note the presence of cracks in the specimen surface at 400 °C that are not in evidence in the specimen tested at 20 °C.

A common difficulty encountered in hot hardness testing is deterioration of the indenter at high temperatures. The traditional diamond indenter graphitizes at temperatures above 1000 °C. Other indenter materials such as alumina, silicon nitride, silicon carbide, and cubic boron nitride offer possibilities but are often unsuitable because their hardness is usually comparable to that of the prospective test specimen. However, the technique of mutual indentation can be used in these circumstances.[22] In mutual indentation testing, two crossed cylinders or wedges are bought into contact and their relative displacement measured.

Nanoindentation testing at elevated temperatures is relatively rare but is of considerable practical importance ranging from the performance of structural materials on a micron scale in the semiconductor industry to the performance of thermal barrier coatings and the surface properties of ceramic turbine blades.

9.6 Strain-Hardening Exponent

Traditional tensile tests are often used to generate a uniaxial stress strain curve for the specimen material from which elastic and plastic behaviour of the specimen material may be measured. For an ideal elastic–plastic material, the stress reaches a maximum value and thereafter remains constant with increasing strain. For many materials (e.g., annealed metals), the stress increases with increasing strain due to strain-hardening. Strain-hardening occurs due to the pile-up and interaction between dislocations in the material. These interactions serve to make the material "harder."

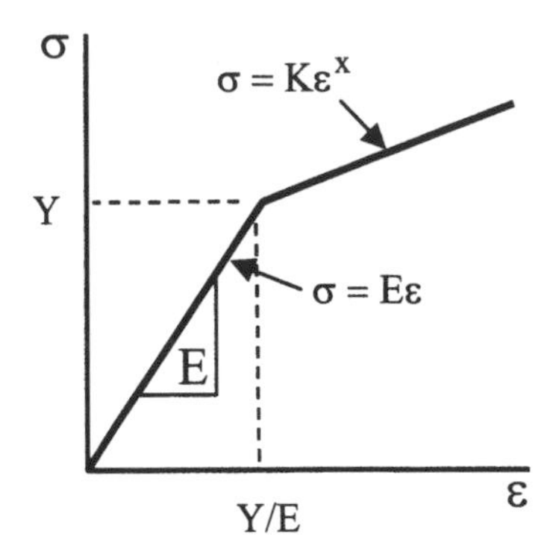

Fig. 9.7 Uniaxial stress–strain curve for an ideal elastic–plastic specimen material with strain-hardening index x. When $x = 0$, the material is elastic perfectly-plastic.

As shown in Chapters 4 and 6, the strain-hardening exponent x is a measure of the strain-hardening properties of a particular material, and this is illustrated in Fig. 9.7. For $x = 0$, the solid is elastic perfectly-plastic. In many practical applications, uniaxial tensile tests are not available, or are unsuitable. A nanoindentation test, on the other hand, is a virtually non-destructive test method that yields values of hardness, modulus, and, in some circumstances, strain-hardening exponent.

There are a number of ways to determine the strain-hardening exponent. One of the most straight forward methods is to obtain an experimental load-displacement curve, using a spherical indenter, and then using the methods of Chapter 5 to create a simulated, or theoretical curve, adjusting E, H, and x to obtain a best fit to the experimental data (see Eq. 5.2).

An alternative method is to make use of Meyer's original work in which it was found that the slope of a plot of log P vs log d, where d is the diameter of the residual impression, gave a value for Meyer's index, by which the strain-hardening exponent could be calculated from Eq. 2.3.1d

Shinohara, Yasuda, Yamada, and Kinoshita[31] use an alternative approach specifically directed towards metals such as copper, aluminum, and nickel. It was found that in copper, the measured hardness H using a pyramidal indenter showed a dependence on the indenter load P and this load dependence systematically depended upon the value of the strain-hardening index x. These workers also showed that the slope of H vs log P was linear and decreased with increasing n. However, such a procedure requires many measurements of H and P over a range of loads, and also requires tensile tests to be performed to establish the specific relationship between x and the slope for the material being tested. To overcome these limitations, a generalized relationship:

$$\frac{H(P)}{H_o(P)} = -0.83x + 0.95\,. \tag{9.6a}$$

was proposed where $H_o(P)$ is the value of H obtained on the metal that shows no strain-hardening — i.e. fully hardened metals.

Ahn and Kwon[32] performed instrumented indentation tests on a macroscopic scale using spherical indenters. Piling-up and sinking-in for metals with a low value of yield stress were accounted for in terms of the strain-hardening exponent according to[4]:

$$\frac{a^2}{a^{*2}} = \frac{5}{2}\left(\frac{2-x}{4+x}\right). \tag{9.6b}$$

where a is the actual contact radius and a^* is the contact radius without correction for piling-up or sinking-in. The real contact radius is given by:

$$a = \frac{5}{2}\left(\frac{2-x}{4+x}\right)\left(2Rh_p^* - h_p^{*2}\right). \tag{9.6c}$$

where h_p^* is the plastic depth without correction for piling-up or sinking-in. Ahn and Kwon determine the unknown value of strain-hardening x using an iterative technique. When the representative strain is given by $\varepsilon = 0.1\cot\alpha$, and piling-up accounted for by Eq. 9.6c, the constraint factor was found to be ≈ 3 for a representative range of steels.

Capehart and Cheng[33], through a finite element study, undertook sensitivity analysis of the general procedure of determining fundamental materials parameters for the case of strain-hardening response from the unloading curve force displacement data for a sharp conical indenter. Unlike the case of an elastic-perfectly plastic contact, the presence of strain-hardening introduces an additional length scale into the problem which, at a 1% noise level in the data, precludes the reliable estimation of E, Y and x from a single unloading curve.

9.7 Impact

Results from conventional quasi-static indentation tests are sometimes not easily correlated with actual product performance where the actual surfaces are subjected to erosive wear, multiple impacts, or stress-strain cycling. A typical example is a tool bit coating which experience an aggressive vibration contact with the workpiece during machining.

Impact testing on a nanometre scale is a fairly novel testing technique. In one commercially available device, a repetitive contact at a single point is obtained by oscillating the specimen against the indenter that is mounted on a freely swinging pendulum. When a change in the contact energy occurs (after fracture of the surface or partial debonding of a coating), a change in indenter recoil takes place. Changes in indentation depth are a measure of surface damage. With scanning impact testing, the specimen is moved sideways at either constant or steadily increasing normal load and this causes the indenter to continuously impact along a wear track, simulating many in-service film failure situations, e.g., erosive wear and coating adhesion failure. The impact energy is determined by the oscillation amplitude, frequency, and the applied load.

For quantitative impact testing, the pendulum is moved away from the specimen by a known distance and then released to produce a single impact. The impact energy is known. Successive impacts can be produced at a single point until failure occurs. In most cases, an initial period of fatigue damage generation occurs, in which small cracks develop and expand, but in which no appreciable increase in penetration depth is observed. Eventually, the cracks coalesce, material is removed, and a sudden depth increase in recorded.

Figure 9.8 shows an impact test result for a 567 nm thick diamond-like carbon (DLC) film on a silicon substrate using the successive single impact method. In this test, there was one impact every 4 seconds, each with an energy of 27 nJ. The indenter was a 25 μm radius diamond sphere. The result illustrates fatigue crack growth (the diamond begins to move away from the surface due to increasing crack volume as evidenced by an decrease in indentation depth) followed by an abrupt film delamination. The silicon substrate itself showed no sign of failure under the same test conditions.

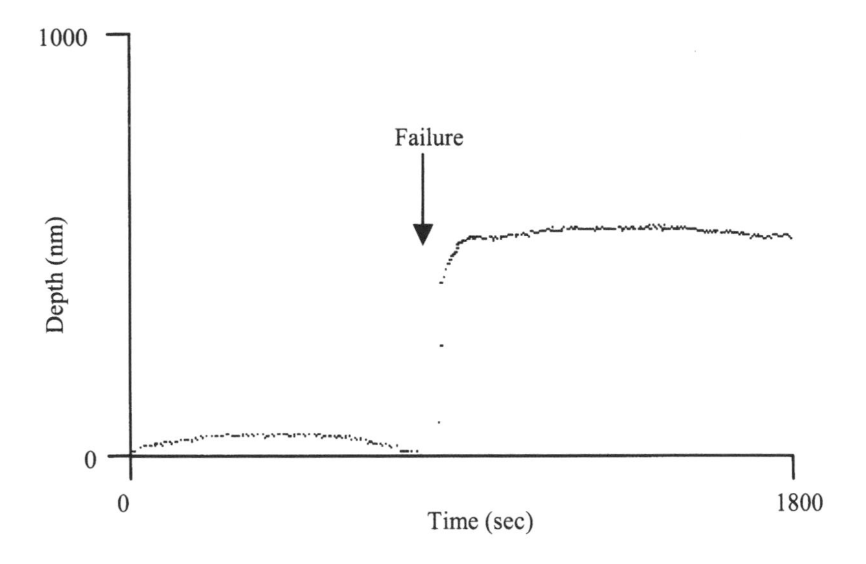

Fig. 9.8 Impact result for 567 nm thick diamond-like carbon (DLC) film on a silicon substrate with 25 μm radius indenter. One impact every four seconds. Failure of the film occurred after 14 minutes (Courtesy Micro Materials Ltd.).

One quantitative result from impact testing is the "dynamic hardness," which is defined as the energy consumed during the indentation divided by the volume of the indentation. The energy of the indentation can be determined from the ratio of the impact and rebound velocities. An example of the technique is given in Chapter 12.

References

1. M. Shiwa, E.R. Weppelmann, D. Munz, M.V. Swain, and T. Kishi, "Acoustic emission and precision force-displacement observations on pointed and spherical indentation of silicon and TiN film on silicon," J. Mat. Sci. 31, 1996, pp. 5985–5991.
2. N.I. Tymiak, A. Daugela, T.J. Wyrobek, and O.L. Warren, "Highly localized acoustic emission monitoring of nanoscale indentation contacts," J. Mater. Res. 18 4, 2003, pp. 784–796.
3. D. Tabor, *Hardness of Metals*, Clarendon Press, Oxford, 1951.
4. R. Hill, B. Storåkers, and A.B. Zdunek, "A theoretical study of the Brinell hardness test," Proc. Roy. Soc. A423, 1989, pp. 301–330.
5. M.M. Chaudhri, "Subsurface deformation patterns around indentation in work-hardened mild steel," Phil. Mag. Lett. 67 2, 1993, pp. 107–115.
6. A.F. Bower, N.A. Fleck, A. Needleman, and N. Ogbonna, "Indentation of a power-law creeping solid," Proc. Roy. Soc. A441, 1993, pp. 97–124.
7. B. Storåkers and P. -L. Larsson, "On Brinell and Boussinesq indentation of creeping solids," J. Mech. Phys. Solids, 42 2, 1994, pp. 307–332.
8. M.J. Mayo and W.D. Nix, "A microindentation study of superplasticity in Pb, Sn, and Sn-38wt%Pb," Acta Metall. 36 8, 1988, pp. 2183–2192.
9. N.R. Moody, A. Strojny, D. Medlin, S. Guthrie, and W.W. Gerberich, "Test rate effects on the mechanical behaviour of thin aluminium films," Mat. Res. Soc. Symp. Proc. 522, 1998, pp. 281–286.
10. Y.-T. Cheng and C.-M. Cheng, "What is indentation hardness?," Surf. Coat. Tech. 133-134, 2000, pp. 417–424.
11. C.A. Schuh, T.G. Nieh and Y. Kawamura, "Rate dependence of serrated flow during nanoindentation of a bulk metallic glass," J. Mater. Res. 17 7, 2002, pp. 1651–1654.
12. N.Q. Chinh, Gy. Horváth, Zs. Kovács, J. Lendvai, Characterization of plastic instability steps occurring in depth-sensing indentation tests," Mat. Sci. and Eng. A324, 2002 pp.219–224.
13. A.C. Fischer-Cripps, *Introduction to Contact Mechanics*, Springer-Verlag, New York, 2000.
14. S. Palmqvist, "A method to determine the toughness of brittle materials, especially hard materials," Jernkontorets Ann. 141, 1957, pp. 303–307.
15. B.R. Lawn, A.G. Evans, and D.B. Marshall, "Elastic/plastic indentation damage in ceramics: the median/radial crack system," J. Am. Ceram. Soc. 63, 1980, pp. 574–581.

16. G.R. Anstis, P. Chantikul, B.R. Lawn, and D.B. Marshall, "A critical evaluation of indentation techniques for measuring fracture toughness: I Direct crack measurements," J. Am. Ceram. Soc. 64 9, 1981, pp. 533–538.
17. M.T. Laugier, "Palmqvist indentation toughness in WC-Co composites," J. Mater. Sci. Lett. 6, 1987, pp. 897–900.
18. F. Ouchterlony, "Stress intensity factors for the expansion loaded star crack," Eng. Frac. Mechs. 8, 1976, pp. 447–448.
19. R. Dukino and M.V. Swain, "Comparative measurement of indentation fracture toughness with Berkovich and Vickers indenters," J. Am. Ceram. Soc. 75 12, 1992, pp. 3299–3304.
20. J.S. Field, M.V. Swain, J.D. Dukino, "Determination of fracture toughness from the extra penetration produced by indentation pop-in," J. Mater. Res. 18 6, 2003, pp. 1412–1416.
21. E.R. Petty and H. O'Neill, "Hot hardness values in relation to the physical properties of metals," Metallurgica, 63, 1961, pp. 25–30.
22. A.G. Atkins and D. Tabor, "Mutual indentation hardness apparatus for use at very high temperatures," Brit. J. Appl. Phys. 16, 1965, pp. 1015–1021.
23. A.G. Atkins and D. Tabor, "Hardness and deformation properties of solids at very high temperatures," Proc. Roy. Soc. A292, 1966, pp. 441–459.
24. A.G. Atkins and D. Tabor, "The plastic deformation of crossed cylinders and wedges," J. Inst. Metals, 94, 1966, pp. 107–115.
25. E.A. Payzant, H.W. King, S. Das Gupta, and J.K. Jacobs, "Hot hardness of ceramic cutting tools using depth of penetration measurements," in *Development and Applications of Ceramics and New Metal Alloys*, H. Mostaghaci and R.A.L. Drew, eds. Canadian Institute of Mining and Metallurgy, Montreal, 1993.
26. T.R.G. Kutty, C. Ganguly, and D.H. Sastry, "Development of creep curves from hot indentation hardness data," Scripta Materialia, 34 12, 1996, pp. 1833–1838.
27. T. Suzuki and T. Ohmura, "Ultra-microindentation of silicon at elevated temperatures," Phil. Mag. A 74 5, 1996, pp.1073–1084.
28. S.A. Syed Asif and J.B. Pethica, "Nano-scale indentation creep testing at non-ambient temperatures," J. Adhesion, 67, 1998, pp. 153–165.
29. B.D. Beake and J.F. Smith, "High temperature nanoindentation testing of fused silica and other materials," Phil. Mag. A 82 10, 2002, pp. 2179–2186.
30. A.C. Fischer-Cripps and C. Comte, unpublished work.
31. K. Shinohara, K. Yasuda, M. Yamada, and C. Kinoshita, "Universal method for evaluating work-hardening exponent of metals using ultra-microhardness tests," Acta. Metall. Mater. 42 11, 1994, pp. 3909–3915.
32. J.H. Ahn and D. Kwon, "Derivation of plastic stress-strain relationship from ball indentations: Examination of strain definition and pileup effect," J. Mater. Res. 16 11, 2001, pp. 3170–3178.
33. T.W. Capehart and Y.-T. Cheng, "Determining constitutive models from conical indentation: Sensitivity analysis," J. Mater. Res. 18 4, 2003, pp. 827–832.

Chapter 10
Nanoindentation Test Standards

10.1 Nanoindentation Test Standards

The ISO (the International Organization for Standardization) has recently issued a draft international standard ISO 14577 entitled "Metallic materials — Instrumented indentation test for hardness and materials parameters."[1] This draft standard covers depth-sensing indentation testing for indentations in the macro, micro and nano depth ranges and also covers the testing of coated material systems.

The ISO is a worldwide federation of national standards bodies. Preparation of international standards is normally carried out through ISO technical committees. Member bodies interested in a subject for which a technical committee has been established have the right to be represented on that committee.

International standards are prepared in accordance with ISO/IEC Directives, Part 3. Draft international standards are circulated to the member bodies for voting. Publication as a standard requires approval by at least 75% of the member bodies. Draft international standard ISO 14577 was prepared by Technical Committee ISO/TC 164, Mechanical testing of metals, Subcommittee SC 3. In this chapter, the basic features of the draft standard are summarized in some detail.

10.2 ISO 14577

ISO 14577 is (at the time of writing) a proposed international standard that deals with instrumented indentation tests for determining hardness and materials parameters. It is generally agreed that hardness is a measure of a material's resistance to permanent penetration by another harder material. Conventional hardness tests typically use the results of measurements of the size of residual impression made by an indenter loaded onto the specimen surface after the test force has been removed. These types of test ignore any elastic recovery of the specimen material that might occur upon removal of the test load. For large-scale testing, the effects of this are not too severe since experience has shown that the lateral dimensions of most indenters (with the exception perhaps of those made with a Knoop indenter) are not significantly different from full load to full unload.

ISO 14577 allows the evaluation of indentation hardness using instrumented indentation (or depth-sensing indentation) where both the force and displacement during plastic and elastic deformation are measured. Traditional hardness values can be determined as well as indentation hardness and modulus of the test material. With the instrumented indentation technique, it is not necessary to measure the dimensions of the residual impression optically. ISO 14577 consists of four parts together with various annexes.

Part 1 of the Standard contains a description of the method and principles of the indentation test and also contains an annex that provides definitions and methods of calculation of the material parameters to be measured.

Part 2 of the Standard specifies the method of verification and calibration of the test instruments. A direct method is given for verification of the main functions of the instrument and an indirect method for determining the repeatability of the instrument is also given. The annexes to Part 2 give recommendations for the design of the instrument and methods to be used for calibration and verification of the instruments on a periodic basis.

Part 3 of ISO 14577 specifies the method of calibration of reference blocks that are to be used for verification of indentation testing instruments.

Part 4 of the Standard is concerned with the application of the technique to coatings and provides information about test conditions necessary to produce reliable estimations of modulus and hardness for the coating material.

The notation in the standard is different to that presented elsewhere in this book and in some parts of the literature. Figure 10.1 shows the notation used for the relevant features of the indentation test in the standard.

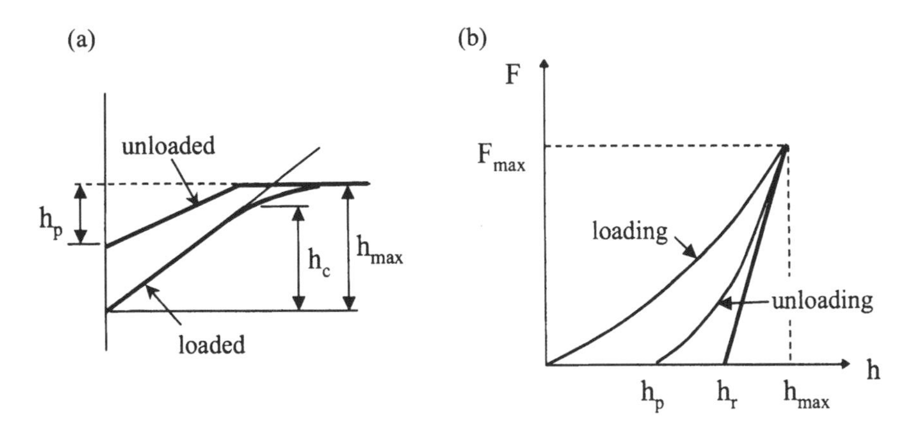

Fig. 10.1 Indentation test procedure. A loading sequence is followed by an unloading sequence. h_p is the depth of the residual impression, h_r is the intercept of the tangent to the initial unloading curve, h_{max} is the maximum penetration beneath the specimen surface, and F_{max} is the maximum load applied to the indenter.

10.2.1 ISO 14577 Part 1: Test method

Part 1 of the standard contains a description of the method and principles of the indentation test and also contains an annex that provides definitions and methods of calculation of the material parameters to be measured.

10.2.1.1 Test Method and Requirements

Part 1 of ISO 14577 specifies the method of instrumented indentation test for determination of hardness and other material parameters for three defined ranges of hardness as shown in Table 10.1, where h is the total indentation depth and F is the test force:

Table 10.1 Ranges of hardness testing as defined in ISO 14577 Part 1.

Macro Range	Micro Range	Nano Range
$2 N \leq F \leq 30\,000 N$	$2 N > F$; $h > 0.02\ \mu m$	$h \leq 0.02\ \mu m$

For testing under this standard, the indenter may take the form of a four-sided Vickers diamond indenter, a Berkovich triangular diamond pyramid indenter,[***] a hardmetal ball of a specified composition (see Section 10.2.2), or a diamond spherical indenter. The Standard allows for the use of indenters with other geometries and manufacture.

The indentation test procedure can either be load or displacement controlled. The test force F, indentation depth h, and time are recorded during the test procedure. The resulting load-displacement data is used to calculate the relevant properties of the specimen material. The initial penetration depths for the displacement measurements are required to be determined for each individual test. For time-dependent effects such as creep and/or thermal drift, the force is to be kept constant and the change of indentation depth is to be recorded (for load-controlled testing) or the depth is to be kept constant by varying the load for depth-controlled testing.

The indentation testing instrument is required to have the capability of applying the test forces within the limits set down in Part 2 of the Standard. The instrument is to have the capability for continuous measuring of applied load, displacement, and time and is required to be able to compensate for the compliance of the instrument and non-ideal indenter geometry by the use of an area function.

The Standard specifies that the indentation test shall be carried out on a specimen surface that is smooth and free from lubricants and contaminants and is of an acceptable surface roughness. The surface of the specimen should be

[***] The original Berkovich indenter has a face angle of 65.03° to give the same actual surface area to depth ratio as a Vickers indenter. The modified Berkovich indenter has a face angle of 65.27° to give the same projected area to depth ratio as a Vickers indenter.

prepared so that its surface properties are not unintentionally modified by cold working or strain-hardening and the specimen thickness shall be great enough so that the results are not influenced by the specimen mounting. As a rule of thumb, the specimen thickness should be at least ten times the indentation depth or three times the indentation diameter.

Each individual test is required to be carried out at stable temperature conditions after both the specimen and instrument have reached an equilibrium temperature and the temperature of the system is to be maintained to within an appropriate tolerance during the test.

During the test, a sufficient number of data points are to be recorded so that initial penetration depth may be calculated with the required accuracy. The initial penetration depth can be calculated using a polynomial regression to the first 10% of the load-displacement data or from the first increase in load during the indenter approach. The load (or displacement) can be controlled either continuously or in steps. Both force and depth are to be recorded at specified time intervals. The applied force should be applied at a constant rate (N/sec) for load controlled testing or to provide a constant displacement rate (mm/sec) for depth controlled testing. The load removal rate is not specified.

Hold periods, where the load is held constant and the depth measured at specified intervals for load controlled testing (or vice-versa in depth control) may be inserted at convenient points in the test cycle. The data taken within the hold periods can be used to determine the thermal drift rate during the test.

Indentations shall be spaced at approximately three to five times their diameter of residual impression so that the results are not affected by the presence of an edge or any previous residual impression in the specimen surface.

The Standard requires an estimation of the uncertainty of the measurement to be made. These arise from uncertainties resulting from the calibration of the instrument, and uncertainties arising from standard deviations from a series of measurements. The Standard also specifies the format of the test report.

10.2.1.2 Analysis Procedures

Part 1 of the Standard deals with the procedure and the principle of the instrumented indentation test. Annex A of the Standard specifies the definitions of the material properties to be calculated and their method of calculation.

10.2.1.2.1 Martens (Universal) Hardness (HM)

The Martens hardness is defined as the test force F divided by the actual surface contact area of the indentation and is measured under applied test force — not from the dimensions of the residual impression. The Martens hardness value HM is defined for Vickers and Berkovich indenters. It is not defined for spherical or Knoop indenters. The Martens hardness was previously designated "Universal hardness" HU. For a Vickers indenter, Martens hardness is given by:

$$HM = \frac{F}{A_s(h)} = \frac{F}{26.43\,h^2} \qquad (10.2.1.2.1a)$$

where

$$A_s(h) = \frac{4\sin(\alpha)}{\cos^2(\alpha)}\,h^2 \qquad (10.2.1.2.1b)$$

For an unmodified Berkovich indenter, the Martens hardness is found from:

$$HM = \frac{F}{A_s(h)} = \frac{F}{26.44\,h^2} \qquad (10.2.1.2.1c)$$

where

$$A_s(h) = \frac{3\sqrt{3}\tan\alpha}{\cos\alpha}\,h^2 \qquad (10.2.1.2.1d)$$

In Eqs. 10.2.1.2.1b and 10.2.1.2.1d α is the face angle of the indenter (68° for a Vickers indenter and 65.03° for a Berkovich indenter) and h is the penetration depth measured from the initial penetration depth. The quantity A_p is the surface area of the indenter that penetrates beyond the initial contact point. Martens hardness values are determined from load and depth readings during the application of the test force. A penetration greater than 0.2 µm depth is required.

The Martens hardness value is denoted by the symbol HM, followed by the test conditions that specify the indenter (if not a Vickers), the test force, the time of application of the test force, and the number of load steps applied if not a continuous application of force. For example, "HM (Berkovich) 0.5/20/30 = 6500 N/mm^2" represents a Martens hardness value of 6500 N/mm^2, determined with a test force of 0.5 N, applied during 20 seconds in 30 steps.

10.2.1.2.2 Martens Hardness HM_S

The Martens hardness can be computed from the slope of the increasing load-displacement curve and, when measured in this way, is designated HM_S. It is found that for many materials, the load and depth are related according to:

$$h = m\sqrt{F} \qquad (10.2.1.2.2a)$$

where m is a constant that depends upon the shape of the indenter which is found by linear regression of data plotted in accordance with Eq. 10.2.1.2.2a. The Martens hardness is then found from:

$$HM_S = \frac{1}{m^2 A_s(h)/h^2} \qquad (10.2.1.2.2b)$$

where $A_s(h)/h^2$ is 26.43 for a Vickers indenter and 26.44 for a Berkovich indenter. This method does not rely on the determination of the initial penetration depth nor is it influenced by surface roughness. However, for specimens that show a variation in hardness as a function of depth, the value determined using this method will be different from that given in Section 10.2.1.2.1.

10.2.1.2.3 Indentation Hardness (H_{IT})

Indentation hardness H_{IT} is defined as the mean contact pressure, that is, the indentation load divided by the projected area of contact, and as such is physically equivalent to the Meyer hardness (see Section 2.3.1). Using the notation specified in the Standard, the indentation hardness is found from:

$$H_{IT} = \frac{F_{max}}{A_p} \tag{10.2.1.2.3a}$$

where F_{max} is the maximum load and A_p is the projected area of contact at that load. A_p is determined from the load-displacement curve. In the Standard, A_p is referred to as the "area function" of the indenter and relates the projected area to the distance from the tip of the indenter. For non-ideal indenters, polynomial fitting, a look-up table or calibration graph, may be used to define the area function. For ideal indenters, the area function can be determined from Table 10.2.

Table 10.2 Projected areas, intercept corrections, and geometry correction factors for various types of indenters used in the determination of Indentation hardness.

Indenter type	Projected area	Geometry correction factor ε
Berkovich	$A_p = 23.97h_c^2$	0.75
Berkovich (modified)	$A_p = 24.5h_c^2$	0.75
Vickers	$A_p = 24.5h_c^2$	0.75

In the Standard, h_c is the depth of contact of the indenter (equivalent to h_p elsewhere in this book) with the specimen given by:

$$h_c = h_{max} - \varepsilon(h_{max} - h_r) \tag{10.2.1.2.3b}$$

where ε is a geometry correction factor given in Table 10.2, h_r is the depth found from extrapolating the slope of the tangent of the initial unloading to the depth axis, and h_{max} is the maximum penetration depth as shown in Fig. 10.1.††† Different methods for the determination of h_r are allowed: A linear fit to the initial unloading data or a power-law fit in accordance with Doerner and Nix[2] and Oliver and Pharr[3] respectively.

The Standard specifies that the upper 80% of the unloading curve is to be taken for the least squares fitting procedure. If only 50% or less of the unloading data are used, the indentation test shall be interpreted with some care. The slope of the tangent is found by differentiating the least squares fitted line, or curve, and evaluating this at F_{max}. The intercept of this tangent with the displacement axis gives a value for h_r.

The indentation hardness H_{IT} value is expressed together with the test conditions in a manner similar to that for the Martens hardness.

10.2.1.2.4 Indentation Modulus (E_{IT})

The indentation modulus E_{IT} is calculated from the slope of the tangent for the calculation of indentation hardness following the method given by Oliver and Pharr.[3] The indentation modulus typically provides values that are similar to Young's modulus for the specimen material (the value may differ from the accepted value due to piling-up or sinking-in during the indentation process). The indentation modulus is found from:

$$E_{IT} = \frac{1-(\nu_s)^2}{\dfrac{1}{E_r} - \dfrac{1-(\nu_i)^2}{E_i}} \qquad (10.2.1.2.4a)$$

where the subscripts s and i refer to properties of the specimen and the indenter, respectively. For the specimen, E_r is the reduced modulus (E^*), which is found from the indentation test data and is given by:

$$E_r = \frac{\sqrt{\pi}}{2C\sqrt{A_p}} \qquad (10.2.1.2.4b)$$

where C is the compliance of the contact, dh/dF, ν_s is Poisson's ratio of the specimen, and A_p is the projected contact area given by:

$$\sqrt{A_p} = 4.950 h_c \qquad (10.2.1.2.4c)$$

for a Vickers and modified Berkovich indenter. For a Berkovich indenter:

$$\sqrt{A_p} = 4.896 h_c \qquad (10.2.1.2.4d)$$

††† Note, the terminology used in the draft standard is slightly different from that used elsewhere in this book. In particular, the geometry correction factor ε given above should not be confused with the geometry correction factor β described in Chapter 3.

The indentation modulus is expressed together with the test conditions in the following manner: E_{IT} 0.5/10 = 210000 N/mm^2. In this example, the indentation modulus is 210000 N/mm^2, determined using a maximum applied test force of 0.5 N, which was removed continuously over a period of 10 seconds. Interestingly, no correction is applied to account for the non-axial-symmetric nature of the indenter (β in Eq. 3.2.4.1f). The Standard does refer to the existence of such a correction, but does not require it to be applied. It is very important to note that the Standard makes no representation as to the effects of piling-up and sinking-in of the test surface around the indentation. These phenomena can lead to estimations of modulus and hardness that are not consistent with those expected from tests on bulk materials. For this reason, the Standard refers to these properties as the "indentation" modulus and the "indentation" hardness. In metals that pile-up, it is common to find that the indentation modulus is some 10 – 20% higher than the elastic modulus determined by other methods.

10.2.1.2.5 Creep

Creep within the specimen can occur under indentation loading and manifests itself as a change of the indentation depth with a constant test force applied. The relative change of the indentation depth is referred to as the creep of the specimen material. Creep is indistinguishable from thermal drift so that care must be taken to interpret the results.

Figure 10.2 shows the type of data obtained from a creep test. The creep value C_{IT} is expressed as a percentage and is calculated from:

$$C_{IT} = \frac{h_2 - h_1}{h_1} 100 \qquad (10.2.1.2.5a)$$

where h_1 is the indentation depth at the test force which is then kept constant, t_1 is the time at which the test force is reached, h_2 is the indentation depth at a later time t_2.

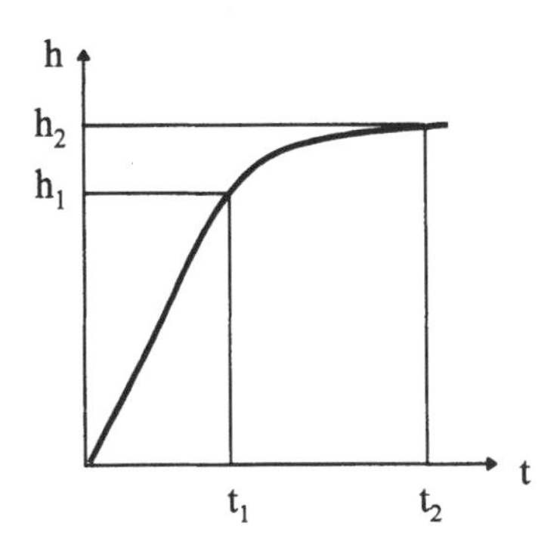

Fig. 10.2 The test force is applied over a time period 0 to t_1. The test force is held constant during time t_1 to t_2 and the change in penetration depth h_1 to h_2 is measured.

The creep value is reported as the relative change of the indentation depth C_{IT} (as a %) together with the test conditions. For example, C_{IT} 0.5/10/50 = 2.5% means a creep of 2.5% determined with a test force of 0.5 N, which was applied in 10 seconds and kept constant for 50 seconds. Note that C_{IT} is *not* expressed as a displacement vs time (mm/sec).

10.2.1.2.6 Relaxation

The relaxation R_{IT} is the relative change of the test force at constant depth. For measuring relaxation, the indentation depth is held constant and the relative change in test force is calculated. The relaxation value is given by:

$$R_{IT} = \frac{F_1 - F_2}{F_1} 100 \qquad (10.2.1.2.6a)$$

and is expressed as a percentage. In Eq. 10.2.1.2.6a, F_1 is the force at the indentation depth. F_2 is the force after the time during which the indentation depth was kept constant. The relaxation value is expressed as the percentage change in force along with the test conditions.

10.2.1.2.7 Indentation Work

The net amount of work done during an indentation test is given by the area enclosed by the load-displacement curve. The ratio of the elastic work recovered during unloading to the total work characterizes the elastic portion of the total work done during an indentation test. This ratio, η_{IT}, is expressed as a percentage and is given by:

$$\eta_{IT} = \frac{W_{Elastic}}{W_{Total}} 100 \qquad (10.2.1.2.7a)$$

where $W_{total} = W_{elastic} + W_{plastic}$. The elastic part of the indentation work is reported as a percentage along with the test conditions. For example, η_{IT} 0.5/10 = 36.5% indicates 36.5% elastic work for a 0.5 N force applied over a time of 10 seconds.

10.2.1.3 Load and Depth Control

Annex B of the Standard provides schematic representations of load and depth control for the different phases in an indentation test: load, creep, and unload.

10.2.1.4 Diamond Indenters

Annex C of the Standard provides information about the use of diamond indenters used in indentation testing. Experience shows that indenters can become defective after a period of use owing to the growth of flaws or cracks. The Standard recommends regrinding indenters that show such deformities upon optical inspection and then recalibrating the indenter.

10.2.1.5 Specimen Roughness

The results of round-robin tests[4] have shown that surface roughness of the specimen can have an influence on the test results. At shallow depths of penetration, asperity contact with the indenter results in relatively large uncertainties in the determination of the contact area. At larger indentation depths, this uncertainty is reduced. In order to obtain an uncertainty of the indentation depth less than 5% of the indentation depth, the indentation depth should be made at least 20 times the arithmetic roughness R_a of the specimen surface.

For example, for a specimen of aluminum, the Standard shows that for a Martens hardness of 600 N/mm^2, the allowed arithmetic roughness for a 0.1 N load is 0.13 μm. For tests in the nano range, it may not be possible to meet the condition of surface roughness for specimens of high hardness. In this case, the number of tests should be increased and this stated in the test report.

10.2.1.6 Instrument Compliance and Indenter Area Function

Annex E of the Standard refers to procedures for correcting the results of the indentation test for the compliance of the load frame of the testing instrument and the non-ideal shape of the indenter.

10.2.1.6.1 Instrument Compliance

As load is applied to the indenter and specimen, reaction forces cause the load frame to be elastically deflected and, in most indentation test instruments, this results in an error in the reported penetration depths. The elastic deformation of the load frame is usually directly proportional to the applied load. Annex E of the Standard recommends that the recorded penetration depths be corrected for this deflection using a method specified in Part 2 of the Standard. The Standard places the responsibility providing a means of establishing the instrument compliance with the manufacturer prior to delivery of the instrument.

10.2.1.6.2 Indenter Area Function

The calculations given in Annex A of the Standard are based on the contact area (or projected contact area) of the indenter with the specimen. However, non-ideal geometry of the indenter, such as blunting of the tip, can cause errors in the estimation of the contact area, especially at small penetration depths. For a Vickers indenter, the line of junction at the tip may also cause errors.

The Standard specifies that the actual area function of the indenter be established for use in calculations. The area function is that function which provides the true contact area as a function of the contact depth h_c. There are three recommended methods of determining the area function:

1. A direct measurement method using an atomic force microscope.
2. Indirectly by performing indentations into a material of known Young's modulus and using the known modulus to determine the true contact area by applying the analysis procedures in reverse.
3. Indirectly by measuring the difference in hardness calculated with the test force and depth to the constant value of hardness for a special reference material that shows no depth-dependent hardness value.

The area function is normally expressed as a mathematical function relating the contact area to the distance from the tip of the indenter. A procedure for the verification of the indenter area function is given in Part 2 of the Standard.

10.2.1.7 Correlation of H_{IT} with Other Scales

The indentation hardness H_{IT} may be correlated to other hardness scales in certain circumstances. For example, a common request is to express indentation hardness as equivalent Vickers hardness value. The indentation hardness uses the projected area of contact, while the Vickers hardness uses the actual surface area of contact. Since for a Vickers indenter, the projected and actual surface areas of contact differ by about 7%, it is to be expected that the Vickers hardness value will be some 7% less than the equivalent indentation hardness H_{IT}. Note that such a conversion assumes perfect indenter geometry which is generally not the case for very small penetration depths.

Annex F of the Standard provides information about the relationship between indentation hardness and Vickers and Berkovich indenters.

10.2.2 ISO 14577 Part 2: Verification and Calibration of Machines

Part 2 of the Standard specifies the method of verification and calibration of the test instruments. A direct method is given for verification of the main functions of the instrument and an indirect method for determining the repeatability of the instrument is also given.

Before verification and calibration, the requirements and guidelines of the manufacturer are to be taken into account when installing the instrument. The test force shall be applied and removed without shock or vibration, and the process of increasing, holding, and removal of the test force be verified.

Direct verification involves verification of the indenter, calibration of the test force, calibration of the displacement measuring device, verification of the machine compliance, verification of the indenter area function, and verification of the testing cycle.

10.2.2.1 Indenters

The indenters used for the indentation tests are required to be calibrated independently of the indentation instrument by a direct optical method, and the calibration certificate must include the relevant geometrical measurements. If the measured angle of the indenter deviates from its nominal value, then the average of the measured angles are to be used in all calculations.

10.2.2.1.1 Vickers Indenter

Vickers indenters have four faces with the angle between the opposite faces of the vertex of the pyramid to be 136 ± 0.3°. The angle between the axis of the diamond pyramid and the axis of the indenter holder is not to exceed 0.5°. The line of conjunction at the tip of the indenter shall be no greater than 1 μm for an indentation depth range > 30 μm, 500 nm for an indentation depth range of 30 to 6 μm, and less than 500 nm for an indentation depth range less than 6 μm. The radius of the tip of the indenter shall not exceed 500 nm for the micro range.

10.2.2.1.2 Berkovich and Cube Corner Indenters

The Standard specifies that the radius of the tip of a Berkovich indenter shall not exceed 500 nm for the micro range and 200 nm for the nano range. The angle between the three faces of the diamond pyramid at the base is specified to be 60° ± 0.3°. The included face angle for the different types of indenter are to be as shown in Table 10.3:

Table 10.3 Face angles for triangular pyramid indenters

Berkovich indenter	65.03° ± 0.3°
Modified Berkovich indenter	65.27° ± 0.3°
Cube corner indenter	35.26° ± 0.3°

10.2.2.1.3 Spherical Indenters

The Standard describes the requirements for metal spherical indenters and diamond sphero-conical indenters. For spherical indenters the hardness shall be not less than 1500 HV 10, when determined in accordance with ISO 3878. The indenters are typically made from tungsten carbide with 5 – 7% cobalt and up to 2% other carbides.

The calibration certificate for the indenter shows the diameter of the average value of at least three measured points of different positions which fall within a specified tolerance to be acceptable for use.

For sphero-conical indenters, with a cone of semi-angle α, the depth at which the spherical tip is defined is given by Eq. 4.9a. In practice, owing to the gradual transition between the sphere and the cone, and allowed tolerances in the dimensions of α and R, the penetration depth should be no greater than 0.5 h_s.

10.2.2.2 Calibration of Force and Depth

The Standard requires that each range of load offered by the instrument be calibrated using at least 16 evenly spaced points and that the procedure be repeated three times. The test force shall be measured by a traceable method using, for example: an elastic proving ring, a calibrated mass, or an electronic balance with an accuracy of 0.1% of maximum test force. The smallest indentation depth to be measured for the micro range is 0.2 μm and for the macro range 2 μm. The displacement measuring device shall be calibrated for every range offered by the instrument using a minimum of 16 evenly distributed points in each direction.

10.2.2.3 Verification of Compliance and Area Function

Verification of instrument compliance shall be made after calibration of the load and depth measurement systems. Instrument compliance is determined by the measurement of hardness or modulus at a minimum of five different test forces on a reference specimen with certified hardness value or modulus.

Procedures for determination of indenter area function are given in Annex D Part 2 of the Standard. This part of the Standard is concerned with verification of the area function only. The verification procedure of the indenter area function consists of a comparison of the measured indenter area function with that determined for the newly certified and calibrated indenter. If the difference of these values at the same test forces exceeds 30% of the initial value, the indenter should be discarded.

10.2.2.4 Verification of the Instrument

The Standard specifies that indirect verification should be carried out at least weekly for instruments in the micro and nano ranges using reference blocks calibrated in accordance with Part 3 of the Standard. Two different reference blocks are to be chosen that span the range of normal application of the instrument. The Standard specifies the method of calculating the mean values of results on each reference block and also the standard deviation of these readings.

The Standard specifies that direct verification shall be carried out when the instrument is installed, after its dismantling or relocation, or when the result of an indirect verification is unsatisfactory. A test shall be performed at two different test forces on a specimen of known properties on a daily basis. The results should be recorded on a time chart and, if the results are outside the normal range, an indirect verification should be performed.

The Standard specifies the format of the Verification Report and the Calibration Certificate. In general, the following information is to be reported: reference to the Standard, method of verification (direct and/or indirect), identification of the testing machine, means of verification (reference blocks, elastic proving devices, etc.), test forces, temperature, results, date of verification, and reference to the verification institution.

10.2.2.5 Annexes to Part 2

Annex A of this part of the Standard describes a recommended design for the construction of the indenter holder. The design is intended to minimize the compliance of the holder and to provide a firm mounting for the indenter material.

Annex B of Part 2 of the Standard is similar to that of Annex C of Part 1 and provides information about methods of cleaning of contaminated indenters. The Standard recommends that the condition of indenters should be monitored by visually checking the aspect of the indentation on a reference block, each day the testing machine is used.

Annex C of this part of the Standard gives examples for direct verification of the displacement measuring system using either a laser interferometer, an inductive method, a capacitive method, or a piezo-translator method.

Annex D of Part 2 of the Standard describes procedures for verification of indenter area function. A series of at least 10 different forces shall be chosen to span the range of interest and, for each load, at least 10 indentations shall be made into the reference material and the mean value used to determine A_p. The area function takes the form of a plot of A_p versus indentation contact depth h_c.

Annex E of Part 2 of the Standard shows examples for the documentation of the results of indirect verification in the form of charts by which the performance of a test instrument can be monitored over time.

10.2.3 ISO 14577 Part 3: Calibration of Reference Blocks

Part 3 of ISO 14577 specifies the method of calibration of reference blocks that are to be used for verification and calibration of indentation testing instruments.

The Standard specifies that the reference block shall be specially prepared with a suitable level of homogeneity, stability of structure, and uniformity. The Standard specifies the minimum thickness of reference block for each hardness range. Reference blocks are calibrated using a calibration indentation test instrument. The calibration instrument is required to be calibrated to a traceable standard in terms of its test force, indenter shape, displacement measurement system, and the testing cycle.

The average and standard deviation for hardness measured on the reference block is to be reported. The maximum value of standard deviation is to be less than 2%. Each reference block is required to be marked with the mean of the measured values of hardness and modulus; the name of the manufacturer of the block; a serial number; the name or mark of calibrating agency; the thickness of the block or an identifying mark on the test surface and the year of calibration. The Standard recommends that calibration validity should be limited to 5 years.

10.2.4 ISO 14577 Part 4: Test Method for Coatings

Part 4 of ISO 14577 provides a standard test method for the indentation of coated materials to obtain the properties of the coating, i.e. excluding the influence of the properties of the substrate. It is based upon the research performed in the EC project 'INDICOAT' led by the National Physical Laboratory, UK[5].

The evaluation of materials properties of coatings is generally made difficult by the relatively low thickness of the coating with respect to the substrate as well as the relative modulus and hardness of the coating and substrate materials. In many material systems of interest, the coating is very thin (on the order of microns or less) and therefore, instrumented indentation testing using nanoindentation test instruments is usually required. It is not possible to obtain the properties of the coating by an indentation method without them being influenced by the elastic properties of the substrate. This part of the Standard seeks to provide a robust procedure for evaluation of indentation modulus and hardness where the influence from the substrate is minimized.

Part 4 of the Standard also provides guidelines for specimen preparation such as allowable surface roughness, recommended cleaning methods, and polishing techniques. These guidelines are similar to those specified in Part 1. The annexes to Part 4 give details of determination of instrument compliance and initial penetration or zero point.

The test procedure itself requires the user to make indentations into the coating at a series of indentation depths. The composite (coating plus substrate) indentation results are then plotted as a function of the normalized parameter a/t_c, where a is the radius of the projected area of contact and t_c is the thickness of the coating. Experiments show that, when such a normalization is performed, composite values for indentation modulus E_{IT} and hardness H_{IT} lie upon an approximately universal curve for a particular material system.

A straight line through the data provides an intercept at $a/t_c = 0$ which is the best estimate of the 'coating only' property value. For coating modulus determination, this approach (within defined ranges of a/t_c) is sufficient and is similar to that shown in Fig. 7.4 (b). However, the Standard recognizes that the indentation response is very different for a hard coating on a soft substrate compared to that of a soft coating on a hard substrate. The Standard places different limitations on the range of normalized depths and the choice of indenter geometry etc. for different types of coating and substrate property combinations.

In the case of a soft coating, the standard requires that the depth range over which indentation tests are performed and subsequently analyzed, be such that $0 < a/t_c < 1.5$. Experimental evidence is given that, in this region, a linear fit of the composite indentation modulus E_{IT}^* vs a/t_c is sufficient to obtain an intercept that it very close to the modulus E_c^* of the coating only. For the case of a hard coating, the suggested range for a linear fit is $0 < a/t_c < 2$.

It is important to note that the Standard only defines indentation hardness, H_{IT}, for geometrically similar indenters as, in such cases, it may be assumed that the strain within the material is constant. The strain within a material loaded with a spherical indenter is not constant with increasing indentation depth but depends upon the ratio a/R. For hardness determinations, the same normalizing parameter a/t_c may be used but for geometrically similar indenters, the ratio h/t_c can also be used.

For hardness determinations of a soft coating on a harder substrate, the coating hardness H_c is obtained from a linear extrapolation to zero of the indentation hardness H_{IT} vs h/t_c. When determining the coating hardness for soft coatings on harder substrates, it is important that the indentation stress does not cause yield within the substrate material.

For the case of a hard coating on a softer substrate, experimental evidence is given that there is an initial rise in hardness value with increasing h/t_c (corresponding to an initial elastic response due to tip rounding) followed by a plateau region within which the hardness is a constant value. This is followed by a reduction in hardness corresponding to yield within the substrate. The Standard recommends that the hardness value of the coating be that obtained within the plateau region. If there is no observable plateau region, then the hardness so obtained is a minimum value of hardness for the coating.

It is important to note that the Standard makes no recommendation as to the validity or usefulness of the commonly used 10% rule other than to say that it is not applicable to modulus measurement and that, in many circumstances, it does not provide a robust method for determining the hardness and modulus of coatings.

A particularly useful feature of Part 4 of the Standard is a series of flowcharts, adapted from Jennett and Bushby[6], that provide a step-by-step decision tree to aid the user in making the right choices when following the testing procedure. For hardness measurements, it is essential to use a sharp-pointed indenter so as to induce full-plasticity at the lowest possible indentation load and depth so that the plasticity is within the coating and occurs before the substrate yield stress is exceeded. A simple Hertzian stress analysis is recommended, using an estimate of the indenter tip radius, to estimate whether the maximum shear stress will fall within the coating rather than the substrate. The hardness decision tree is described in Fig. 10.3. For modulus measurements, it is necessary to consider whether the coating will crack (hard coatings) or creep (soft coatings) under indentation. Using a spherical tipped indenter reduces both of these effects. However, as the radius of the indenter is increased the obtainable values of a/t_c tend to increase.

For modulus measurements, the recommended choice of indenter depends upon the nature of the film and substrate. The decision tree guiding the user in the test procedure is given in Fig. 10.4.

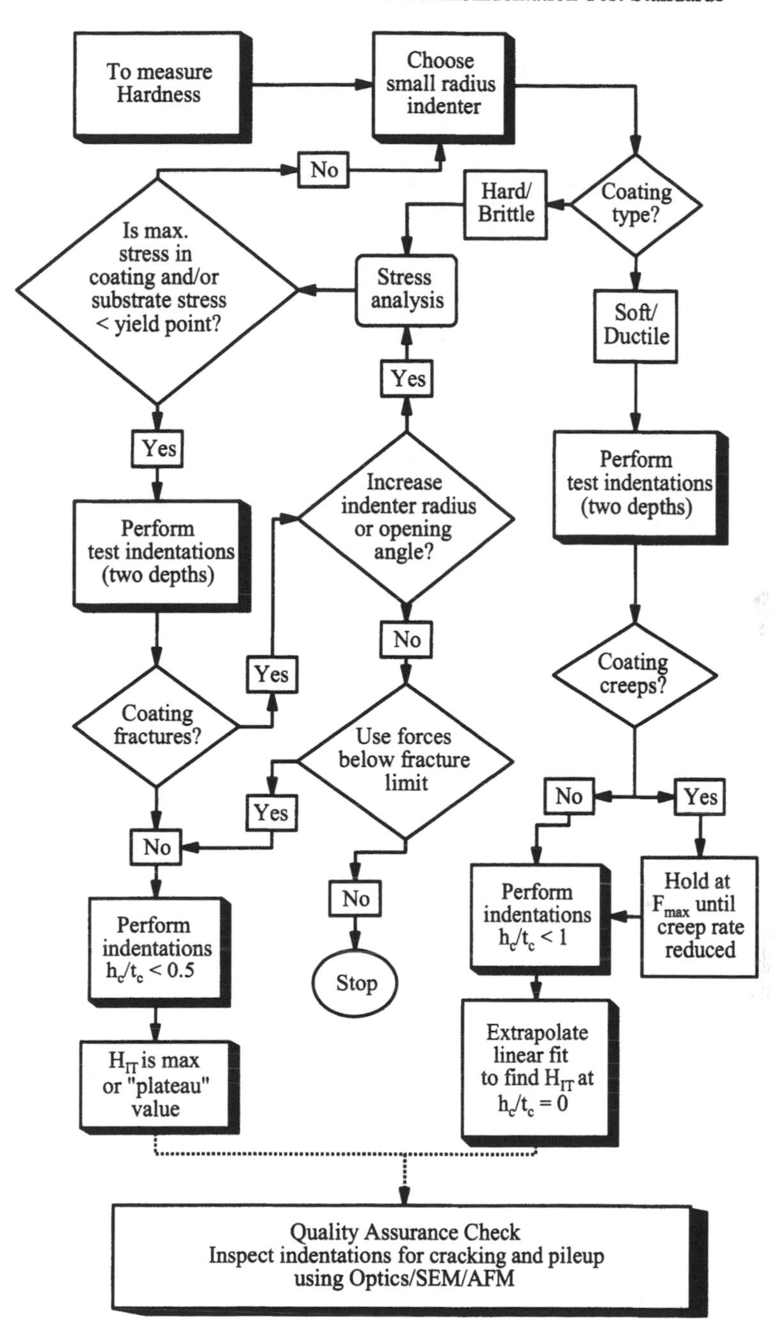

Fig. 10.3 Decision tree to guide the user in the procedure for measuring hardness of a coating on a substrate. (© Crown Copyright 2003. Reproduced by permission of the Controller of HMSO).

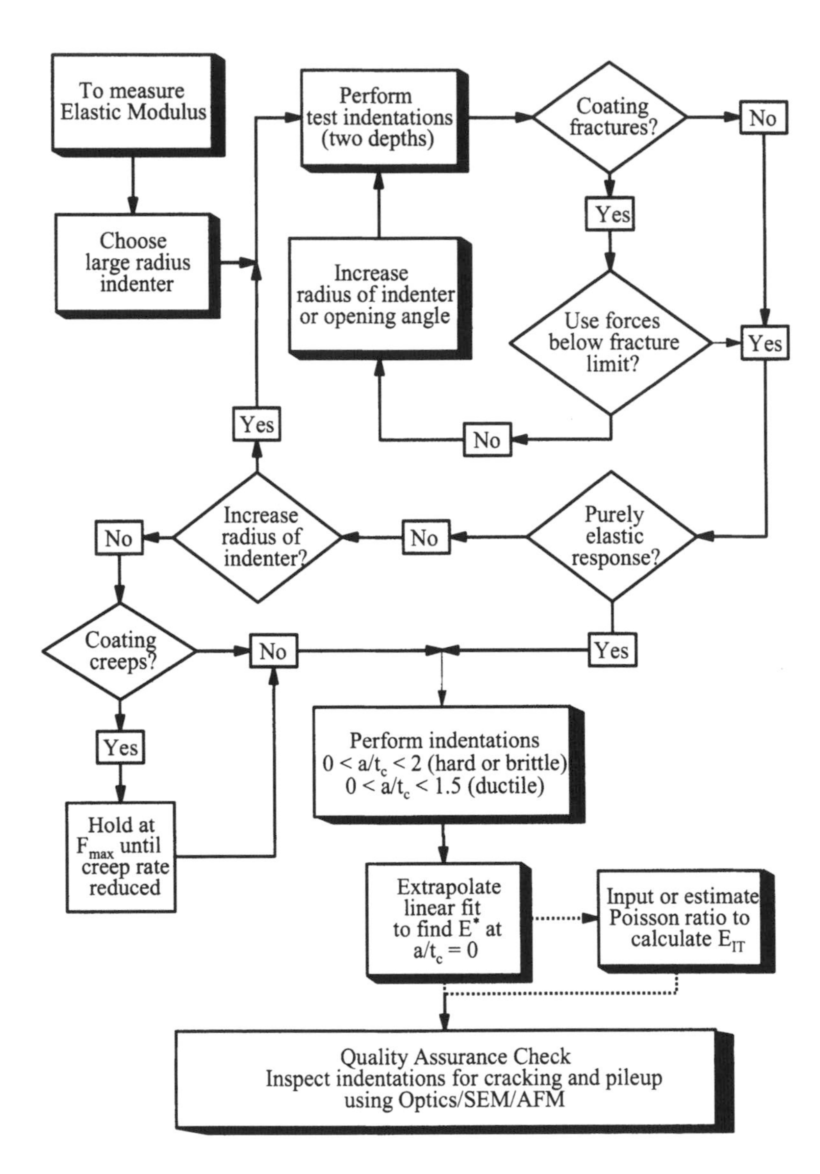

Fig. 10.4 Decision tree guiding the user in the procedure for measuring the elastic modulus of a coating on a substrate. (© Crown Copyright 2003. Reproduced by permission of the Controller of HMSO)

References

1. ISO Central Secretariat, 1 rue de Varembé, 1211 Geneva 20 Switzerland.
2. M.F. Doerner and W.D. Nix, "A method for interpreting the data from depth-sensing indentation instruments," J. Mater. Res. 1 4, 1986, pp. 601–609.
3. W.C. Oliver and G.M. Pharr, "An improved technique for determining hardness and elastic modulus using load and displacement sensing indentation experiments," J. Mater. Res. 7 4, 1992, pp. 1564–1583.
4. H.-H. Behncke, "Bestimmung der Universalhärte und anderer Kennwertr an dünnen Schichten, insbesondere Hartstoffschicten," Härterei-Technische Mitteilung HTM, 48 5, 1993, pp. 3–10.
5. INDICOAT final report, NPL Report MATC (A) 24, NPL, Teddington UK, 2001.
6. N.M. Jennett and A.J. Bushby, "Adaptive Protocol for Robust Estimates of Coatings Properties by Nanoindentation," Mat. Res. Soc. Symp. Proc. 695, 2002, pp.73–78.

Chapter 11
Nanoindentation Test Instruments

11.1 Specifications of Nanoindentation Test Instruments

Interest in nanoindentation has spawned a number of nanoindentation instruments that compete on a world market. Purchasers of such instruments are universities, private and government research organisations, and quality control laboratories. There is particular interest within the semiconductor industry that is concerned with the mechanical properties of a wide range of thin films. All of the products described in this chapter are depth-sensing devices. The instruments typically measure depth of penetration using either an inductance or capacitance displacement sensor. A typical nanoindentation test instrument, or "nanoindenter", has a depth resolution of less than a tenth of a nanometre and a force resolution of several nanonewtons. Load can be applied by the expansion of the piezoelectric element, the movement of a coil in a magnetic field, or electrostatically. Maximum loads are usually limited to the millinewton range. The minimum load is usually less than a micronewton.

Nanoindentation instruments are typically load-controlled machines. A common question asked by the novice is the specification of the minimum thickness of film or specimen that can be measured. This is difficult to answer since it is the minimum load that is the important parameter. When operated at the minimum load, the resulting depth depends upon the mechanical properties of the specimen. The minimum load quoted in manufacturer's specifications is very important and gives an indication of the minimum load range for testing on actual specimens. Force and displacement resolutions are not so important, since they are limited in practice by the noise floor of the instrument and, more importantly, the mechanical, electrical and thermal environment in which it is placed.

Table 11.1 provides a description of the most commonly quoted specifications for nanoindentation instruments. The description of the instruments given in this chapter represents no specific order in terms of market share, price, or number of features nor does it include all the instruments presently available. Details of the specifications and features were obtained from publicly available advertising material and scientific literature or directly from the manufacturer. The prospective purchaser should consult the most up-to-date material provided by the manufacturer, since specifications and features are continually being modified and improved.

Table 11.1 Useful definitions for specifications of nanoindentation instruments.

Minimum Contact Force	The minimum contact force is typically limited by the noise floor of the instrument and the test environment. The value should be as low as possible so as to minimize the error associated with the initial penetration.
Force Resolution	The force resolution determines the minimum change in force that can be detected by the instrument. Most manufacturers would either employ a 16 bit, or 20 bit analogue to digital converter (ADC) in their systems and the theoretical resolution for each instrument can be determined by dividing the range (whether force or depth) by 2 raised to the power of the width of the ADC. For example, for a range of 50 mN and a 16 bit ADC, the theoretical resolution would be 50 mN divided by 2^{16} = 750 nN. Some manufacturers then further divide this value by a factor equal to the square root of the number of readings taken for averaging. The very low values of resolution presented by some manufacturers is thus a combination of the smoothing effect of taking many readings and averaging the results and the width of the ADC and the range. The theoretical resolution is often not the most appropriate measure of performance of an instrument.
Force Noise Floor	The noise floor of the specifications is the most important factor that determines the minimum contact force attainable by the system. Any increase in resolution beyond the noise floor will only mean that the noise is being measured more precisely. The noise floor is generally limited by electronic noise or the environment in which the instrument is located. Typically, the noise floor quoted by manufacturers represents the best possible results obtained under ideal laboratory conditions.
Displacement Resolution	The displacement resolution is typically found by dividing the maximum displacement voltage reading by the number of bits in the data acquisition system.
Displacement Noise Floor	The noise floor in the displacement measurement system will determine what the minimum useable indentation depth. The displacement noise floor is one of the most important measures of performance of an instrument.
Maximum Number of Data Points	This is the maximum number of data points that can be collected for a single test. More data points allow for better resolution of "pop-in" events and other features in the force-displacement curves. However, the data acquisition rate will also be important for large data sets, since data

	should be collected as quickly as possible so as to minimize errors due to thermal drift.
Data Acquisition Rate	This is how fast the machine will collect force and displacement data. The data acquisition rate should be as high as possible so as to allow the time for a test to be shorter, thus minimising errors due to thermal drift.
Variable Loading Rate	The mechanical properties of some specimen materials depend upon the rate of application of load. This ability of vary the loading rate allows such studies to be performed. Slow loading may be desirable for some materials followed by a fast unloading.
Unattended Operation	This is the ability of the instrument to be programmed to collect data on a single site, or an array of sites, while not requiring any operator intervention during the tests.
Specimen Positioning	This is how accurately the instrument can position the indenter. Most instruments allow ± 0.5 μm positioning resolution with optical rotary encoders with some allowing ± 0.1 μm with linear track encoders.
Field of Testing	This is the dimension of the testing area accessible by the indenter based upon maximum movement of the positioning stages. This can be important for allowing tests to be performed on large specimens such as silicon wafers.
Resonant Frequency	This is the natural resonant frequency of the instrument. It depends upon the mass of the instrument and the characteristics of the mounting springs and dampers. A high resonant frequency makes the instrument less susceptible to mechanical environmental interference. Also, a higher resonance allows higher-frequency dynamic measurements to be made. High frequency dynamic measurements can be made with a low resonant frequency system if the specimen, rather than the indenter, is oscillated.
Thermal Drift	Thermal drift is almost unavoidable if the temperature of the environment surrounding the instrument is not kept within very tightly controlled limits. Most instruments are supplied with an enclosure with very high thermal insulating properties.
Machine Stabilisation Time	This is the time needed by the instrument to stabilize after initial power up. The time is usually dependent on the thermal characteristics of the measurement system. A short time allows more efficient use of the test facility.
Indentation Time	This is the average time of a typical indentation cycle from load to full unload. A nanoindentation instrument should be able to perform a single indentation within one or two minutes.

Tip Exchange Time	This is the time needed for the operator to change the indenter. This time should be no more than a minute or so for maximum convenience of operation of the instrument where different indenters are used for different types of testing.
Loading Step	Load can be applied in a variety of ways in typical nanoindentation instruments. A square root spacing of load increments gives an approximate even spacing of depth measurements. A linear spacing of load increments may provide a constant loading rate. The instrument should offer one or two options in this regard.
Constant Strain Rate	Constant strain rate testing involves the application of load so that the depth measurements follow a pre-defined relationship. Some nanoindentation instruments offer the ability of set the loading rate so as to give a constant rate of strain within the specimen material. This may be important for viscoelastic materials or those that exhibit creep.
Topographical Imaging	In-situ topographical imaging provides scans of a surface before it is indented for accurate tip placement and also provides immediate imaging after the indent is completed to measure the size of the residual impression. Such imaging can be done with an atomic force microscope accessory mounted as either another testing station on the instrument assembly or as an in-situ device.
Dynamic Properties	This is the ability to measure the response of surfaces under a sinusoidal or other oscillating load. This technique is important for measuring the viscoelastic properties of materials. The method usually involves the application of an oscillatory motion to the indenter or the specimen. A lock-in amplifier measures phase and amplitude of force and displacement signals.
High Temperature Testing	It is sometimes of considerable interest to measure the mechanical properties of materials at their operating temperature. Some nanoindentation instruments allow testing of smaller sized specimens at temperatures ranging from –5 to +500 °C.
Acoustic Emission Testing	This allows the fitting of an acoustic microphone to the indenter or specimen for recording of non-linear events such as cracking or delamination of thin films.

11.2 "Nano Indenter®," MTS Systems Corporation[1]

The Nano Indenter® indentation instrument has a development history dating back to about 1981.[2] The instrument applies load via a calibrated electromagnetic coil and displacement of the indenter is measured using a capacitive plate transducer.

The load and displacement resolutions are reported to be 50 nN and 0.04 nm respectively.[3] The continuous stiffness measurement[4], "CSM," option is of particular interest in this instrument. The analysis methods given in Chapter 3 show how a measurement of the stiffness of the contact (dP/dh) between the indenter and the materials being tested can be used in the multiple-point unload technique to determine the elastic modulus and hardness of the specimen material. This can be done by partially unloading the indenter at each load increment or by superimposing a small sinusoidal load signal onto the normal load signal. With the CSM technique, the latter method is used. The CSM method has an added benefit in that if the specimen material has a viscoelastic behaviour, then the phase difference between the force and depth signals provides information about the storage and loss moduli of the specimen.

The Nano Indenter® can operate in a scratch testing mode that is also suitable for surface profile measurements. The optional lateral force measurement system provides a friction force measurement capability.

The Dynamic Contact Module (DCM) is a low-load accessory for the Nano Indenter®.[3] Its operation is similar in principle to the standard Nano Indenter® system but offers a high resonant frequency, an increased dynamic frequency range, and a low damping coefficient. This makes the unit less sensitive to environmental noise than conventional instruments, and with a theoretical displacement resolution of 0.0002 nm and a load resolution of 1 nN, the DCM is suitable for detecting surface forces on an atomic scale.

The Nano Indenter® is operated by the MTS TestWorks® instrument control environment, which is common to all MTS test equipment and allows the user flexibility in the specification of test procedure and data analysis. The TestWorks® software is available in the "Professional" or more advanced "Explorer" level. Included with the installation is the TestWorks® "Analyst" package. This package offers calculations of hardness and modulus (Oliver and Pharr method), calculation of hardness and modulus as a function of depth and the storage and loss moduli (if the CSM option is included). The software also provides flexibility for the user to specify their own calculation methods.

Additionally, the optional NanoSP1® software acts as a user-friendly interface to the finite element-analysis engine, COSMOS®.[5] NanoSP1® software dramatically simplifies the setup and interpretation of simulations of indentation experiments, including those of thin films on substrates. Mesh generation is automated and optimized to produce accurate results with minimal run times on the order of 20 minutes.

11.3 "NanoTest®," Micro Materials Ltd.[6]

The Micro Materials NanoTest® platform has been designed to support three modules: (i) nanoindentation, (ii) scanning for scratch testing, and (iii) impact (for thin film adhesion failure, erosive wear, and contact fatigue).

In the indentation module, a very small, calibrated diamond probe is brought into contact with the specimen surface and load is applied by means of a coil and magnet located at the top of the pendulum. The pendulum is supported by on a frictionless spring flexure. The resultant displacement of the probe into the surface is monitored with a sensitive capacitive transducer and displayed in real time as a function of load.

In the NanoTest® scanning module, the specimen is moved perpendicularly to the axis of the indenter movement allowing either single or repetitive scratch tests. It is important to note that the pendulum spring support is extremely stiff in the scanning direction, thus minimizing errors due to tilt of the loading head as the scratch load is increased. Accurate repositioning combined with optional software enables complex multi-pass scratch tests to be scheduled.

For quantitative impact testing using the impact module, a static load is applied to the pendulum, which is then pushed away from the specimen by a known amount and released, causing the test probe to impact the surface. The impact energy is determined for individual or repetitive impacts.

Figure 11.1 shows a schematic of the relevant features of the instrument including dynamic oscillation for dynamic measurements.

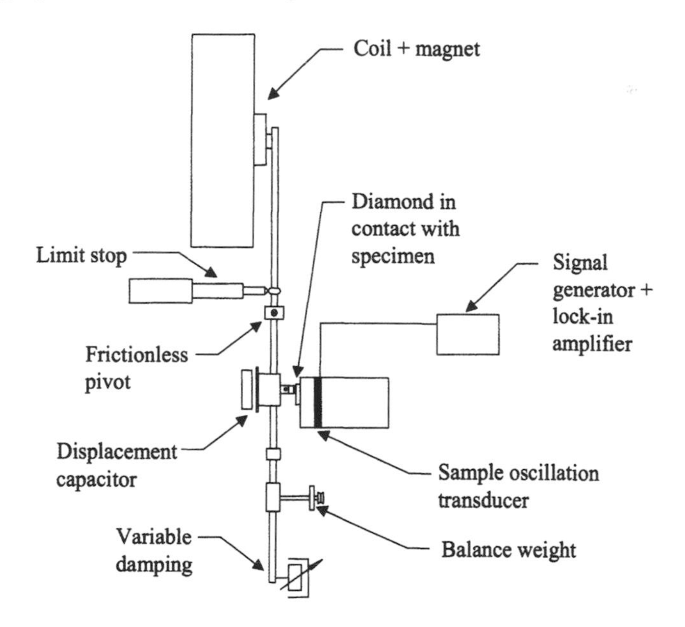

Fig. 11.1 Schematic of the method of construction for the NanoTest® instrument (Courtesy Micro Materials Ltd.).

Significant features and options of the NanoTest® instrument are:

- Unique and versatile pendulum design.
- Precise repositioning technique allows investigation of small particles, fibres, wires, and complex inhomogeneous specimens such as integrated circuits.
- NanoTest® head with load ranges 0 – 50 mN and 0 – 500 mN. Maximum load resolution better than 100 nN. Maximum depth resolution better than 0.1 nm.
- High load capability for load ranges 0 – 2 N and 0 – 20 N. The high load option can be fitted alongside the NanoTest® head.
- Dynamic contact compliance calculation using a lock-in amplifier and specimen oscillation system. The oscillation frequency range is adjustable up to 250 Hz.
- High temperature option offers room temperature to over 500 °C specimen heating stage, probe heater, thermal barrier, and high-temperature capacitor assembly.
- Spherical indenter and analysis software to calculate plastic depth, hardness vs. penetration depth, creep, and stress–strain.
- Automatic 2D specimen leveling stage for hardness/modulus and roughness/topography measurements on curved or uneven specimens.
- High resolution and zoom microscopes with video capture offers high resolution (×1000) with accurate repositioning and specimen translation capabilities. Video zoom 30X – 160X or 60X – 320X. Video capture capability can be added to the system for image storage.
- Pin-on-Disk with rotations of <<1 rpm – 3600 rpm with high-torque motor and gearbox, software for control of speed, acceleration, time, load, depth and track location
- Humidity control system comprising a constant temperature, 15 – 90% RH control unit, designed to operate with a thermally insulated environmental chamber.
- Acoustic emission system with shielded detector, for use with indentation or scanning modules.
- Powder adhesion software to load–unload specimen and measure pull-off forces repeatedly at the same position.
- Integration with an atomic force microscope.
- Fully automated scheduling system — allows overnight operation to ensure maximum productivity — up to 100 nanoindentation experiments each containing up to 100 nanoindentations to be performed at a convenient time at specified locations on one or more specimens.
- Environmental control for excellent data reproducibility even at ultra-low load.

Fig. 11.2 The "NanoTest®" instrument (Courtesy Micro Materials Ltd.).

As shown in Fig. 11.2, there are two load heads side-by-side in the NanoTest® unit. The load head on the left can apply a load of up to 20 N, while the one on the right has a maximum load of 500 mN. In both cases, the maximum depth is 100 μm. Depth resolution is less than 0.1 nm and force resolution is better than 100 nN. The specimen is moved to the front of the microscope for test set-up and then moved to one of the loading heads for actual testing. With the two-head design, it is possible, for example, to perform a wear test with the high-load head and then probe around the wear track with the low-load head.

A particularly unique feature of the NanoTest® instrument is the facility for impact testing. This modular add-on option to the NanoTest® allows the impact technique to be used with or without transverse specimen movement during testing. In one type of test, an oscillating piezoelectric transducer is placed behind the specimen holder. This causes the probe to "bounce" on the specimen surface. The impact frequency, static load, and duration of the experiment are preprogrammed and the impact angle is variable. The initial and final static probe positions are determined in order to calculate the resulting depth increase, and the instantaneous probe position is monitored and plotted throughout the procedure.

With scanning impact testing, transverse specimen movement during scanning at either constant or steadily increasing load causes the test probe to continuously impact along the wear track, simulating many film failure situations, e.g., erosive wear and coating adhesion failure. The impact energy is determined by the oscillation amplitude and frequency and the applied load. Film failure is detected through changes in probe displacement during scanning. The NanoTest® impact testing operates from 0 to 500 Hz and up to 7 μm specimen displacement amplitude.

Another unusual feature offered by the NanoTest® instrument is the optional high-temperature attachment. In this configuration, the displacement measurement capacitor is moved from its original position on the indenter holder to the bottom of the pendulum and a thermal shield is placed between the pendulum and the stage. A very small heater element capable of maintaining a maximum temperature of 500 °C together with a miniature thermocouple have been added to the diamond indenter stub, close to the tip itself. With both the diamond and specimen at the same temperature, heat flow between them does not occur upon contact, thus preventing instantaneous dimensional changes due to thermal expansion. The hot stage itself consists of a thermally insulating ceramic block that is attached to the NanoTest® specimen holder. With the heater at 500 °C, the increase in temperature behind the ceramic block is typically less than 1 °C. Temperature controllers with automatic tuning are used for both the main hot stage and the diamond heater.

The overall distinguishing feature of the NanoTest® system is its versatility, offering conventional nanoindentation (including oscillatory motion), impact, scratch, surface forces, and high-temperature testing.

11.4 "TriboIndenter®" and "Ubi 1™," Hysitron Inc.[7]

The Hysitron TriboIndenter® is a fully automated multi-load range indentation/scratch test system designed for measuring hardness, elastic modulus and dynamic viscoelastic properties of thin films, coatings, bulk materials and MEMS/NEMS type structures. The TriboIndenter® provides quantitative testing capabilities with both normal and lateral force (nano-scratch) loading configurations. It can operate in quasi-static or dynamic loading modes and has AE (acoustic emission) testing capabilities. The TriboIndenter® was developed as an automated, high-throughput instrument that has been built on a single platform designed to support these numerous nanomechanical characterization techniques. The high-performance staging system offers excellent stability and flexibility to accommodate a wide range of applications, sample types, and sizes. The main unit, with the environmental cover removed, is shown in Fig. 11.3.

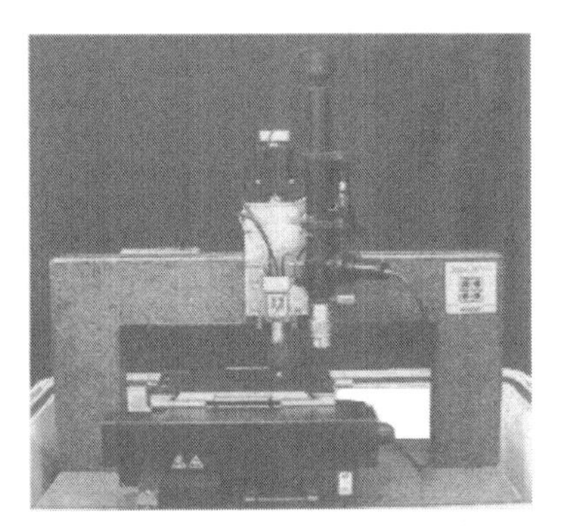

Fig. 11.3 The "TriboIndenter®" (Courtesy Hysitron Inc.).

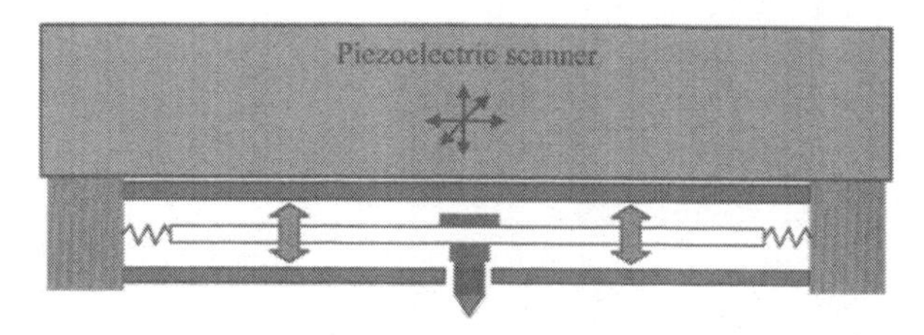

Fig. 11.4 The unique combined sensor/actuator transducer of the Hysitron TriboIndenter® is compatible with in-situ SPM type raster scanning over the specimen surface.

The patented transducer[8–12] used in all Hysitron instruments is based on revolutionary three-plate capacitor technology providing simultaneous actuation and measurement of force and displacement (see Fig. 11.4) with a range of 1 nN to 30 mN. Because of the compact size of the sensor/actuator, it can be interfaced with a piezoelectric scanner that provides very precise X, Y, and Z controlled indenter tip positioning. The piezoelectric scanner is able to raster the tip over a specimen, while a feedback loop controls the Z axis height of the scanner to maintain a constant force between the indenter tip and specimen. The Z axis movement of the scanner is then calibrated to a height to obtain a three-dimensional topographical image. Using in-situ imaging, it is possible to place indentations on a surface with a precision of ±10 nm.

The impressive low-drift characteristics are also a result of Hysitron's transducer design. The TriboIndenter®, when operated in a dynamic mode, can detect forces as small as 1 nN, and respond to phenomena such as surface pull-on and pull-off forces typically associated with an AFM force-depth curve. In quasi-static mode, the TriboIndenter® has a noise floor better than 100 nN. The displacement noise floor for the TriboIndenter® is less than 0.2 nm. This allows the user to make repeatable indentations below 10 nm maximum depth.

The TriboIndenter® can be operated in open loop or closed-loop displacement or force control modes. This means that the user chooses the mode of operation and amount of force or displacement that will be applied with the indenter. Using the Load Function Generator in the TriboIndenter® software, the user can define any type of loading profile. The dynamic stiffness measurement software provided by Hysitron is completely automated for setup, execution, and analysis of dynamic tests and allows measurement of the viscoelastic properties of materials. The lock-in amplifier that measures phase and amplitude of signals is software controlled. The high resonance frequency allows higher-frequency testing to be done as well. The low noise floor allows users to experiment at much lower forces and test real surface properties of specimens.

Hysitron's high load module is a closed-loop piezo actuating, capacitive displacement sensing instrument that can accommodate loads up to 5N and sense lateral force in all directions. The specifications and features of the multi-load range Hysitron TriboIndenter® are given in Table 11.2. Low load and high load options can be integrated in one instrument.

Table 11.2 Specifications and features of the Hysitron TriboIndenter®.

	Low Load Option	**High Load Option**
Force		
Maximum	30 mN	0.5 – 5 N
Minimum Contact	1 nN	5 μN*
Resolution	< 1 nN	< 0.35μN*
Noise Floor	< 100 nN	< 3.5μN*
Displacement		
Maximum	10 μm	80 μm
Resolution	0.0002 nm	0.002 nm
Noise Floor	0.2 nm	2 nm
Positioning		
Placement of Indents	± 10 nm	±250 nm
Field of Testing	120×100 mm/ 12" diameter wafer	120×100 mm / 12" diameter wafer
Instrument Features		
Resonance Frequency	150 Hz	500 Hz
Load Frame Stiffness	0.1×10^7 N/m	0.3×10^7 N/m
User Interface	Windows® 95/98/NT/2000/XP	Windows® 2000/XP
Testing Capabilities		
Constant Loading Rate	50 mN/sec max	0.5 N/sec max
Constant Strain Rate	Yes	Yes
Step Loading	Yes	Yes
In-Situ Topographical Imaging	Yes	Yes
Dynamic Properties	Yes	Yes
Acoustic Emission	Yes	Yes
Temperature Control	–5 °C to +150 °C	–5 °C to +150 °C

The Ubi 1™, a dedicated scanning nanoindenter, has the quasi-static indentation functionality of the TriboIndenter® except for smaller footprint and sample stage (50×50 mm). Its periscope optics allows positioning of the indenter tip close to the sample surface before an indentation test. Automation on Ubi 1™ can be accomplished at two different scales, piezo automation at sub-micron scale and stage automation at micron scale. Thousands of tests can be performed without user intervention. The Ubi 1™ offers high performance desk-top nanoindentation testing for dedicated research applications.

Overall distinguishing features of TriboIndenter® and Ubi 1™ systems are extremely low noise floor allowing very shallow penetration depths of less than

10 nm and piezoelectric scanning allowing in-situ imaging of pre and post indentation features on the specimen surface.

11.5 "Nano-Hardness Tester®," CSM Instruments[13]

In the Nano-Hardness Tester® an indenter tip with a known geometry is driven into a specific site of the material to be tested by applying an increased normal load. Indenter displacement is measured using a capacitive transducer.[14] Using the partial-unload technique, the contact stiffness and hardness of the specimen material can be calculated as a function of depth of penetration into the specimen. The instrument applies load via a calibrated electromagnetic coil and displacement of the indenter is measured using a capacitive sensor. A photograph of the instrument is shown in Fig. 11.5.

A particular feature of this instrument is the use of a sapphire reference ring that remains in contact with the specimen surface during the indentation. The reference ring provides a differential measurement of penetration depth and thus the load frame compliance and thermal drift are automatically compensated for. The sapphire ring also acts as a local environmental enclosure protecting the measurement spot from air currents, sound waves, and changes in humidity and temperature, thus obviating the need for special environmental conditions in order to obtain perfect measurements. The ring also allows the working distance to be set very rapidly as the ring contacts the surface before the indenter and so the final approach, to be made very slowly, takes place only over a few microns displacement.

One of the main advantages of this instrument is the ability to quickly and easily make an AFM image of a residual imprint. As shown in Fig. 11.6 (a), an optional AFM objective fits to the microscope in place of a standard optical objective.

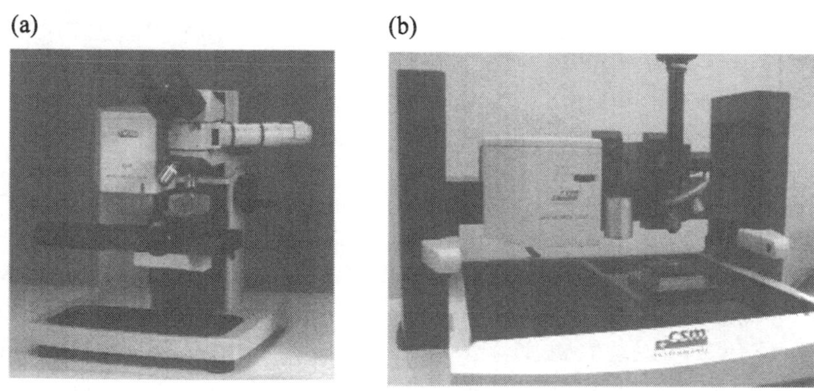

Fig. 11.5 The "Nano-Hardness Tester®" in (a) standard configuration and (b) as part of the Open Platform® (Courtesy CSM Instruments).

The Nano-Hardness Tester® specimen holder will accommodate specimens up to 100 mm thick, on a 240×120 mm work table. Special specimen holders are available to accommodate other specimen sizes such 12 inch silicon wafers. The precise positioning (0.5 μm, 0.1 μm optional) capability of the X-Y motorized table combined with the large X-Y ranges (20×20 μm or 40×40 μm or 80×80 μm) of the objective ensures that the indent will always be in the centre of the field of view in both optical and AFM modes.

Specifications of the Nano-Hardness Tester®:

Depth resolution	0.03 nm
Maximum indentation depth	>20 μm
Maximum load	300 mN
Load resolution	0.04 μN
Work table dimension	240×120 mm
X-Y range	30×21 mm (45 mm optional)
Spatial resolution	0.5 μm, (0.1 μm optional)
Magnification	50X and 1000X
(200X, 500X, and AFM objectives optional)	
AFM resolution x, y, z	< 1 nm
AFM scan range	20×20 μm, (40×40, 80×80 μm optional)
AFM vertical range	2 μm, (4 μm, 8 μm optional)

The "Dynamic Mechanical Analysis" mode of operation uses sine wave loading curves to obtain a more complete analysis of the mechanical properties of viscoelastic materials. Measurements of phase angle and amplitude between the force sine wave and the penetration depth signal produce the storage and loss modulus of the material.

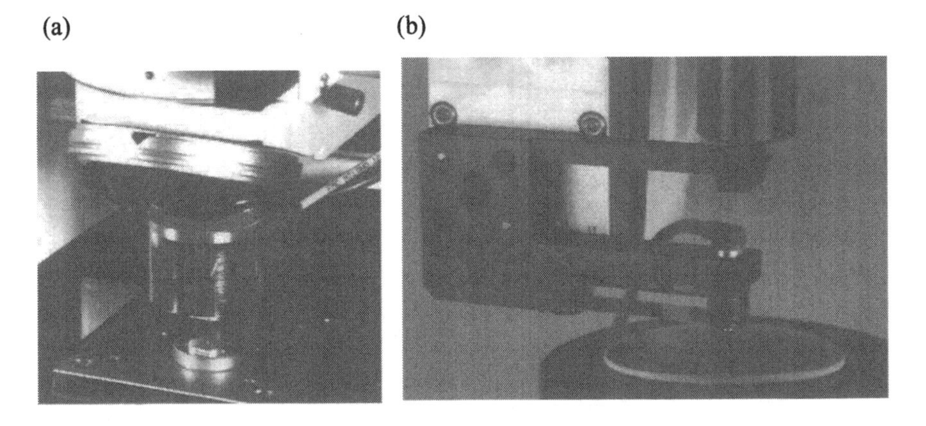

Fig. 11.6 AFM objective on the Nano-Hardness Tester® (a) fits onto the mounting of the standard optical objectives, whilst the Nano Scratch Tester® (b) can be added as an optional head (Courtesy CSM Instruments).

The Nano-Hardness Tester® can be coupled to the Nano-Scratch Tester® and is also available from CSM Instruments and is shown in Fig. 11.6 (b). The Nano-Scratch Tester® operates with a normal force in the range 10 μN – 1 N. Different load ranges are selected by interchangeable cantilevers of varying stiffness. The scratch tip is mounted on the cantilever and the deflection is measured by a linear voltage differential transformer (LVDT). The scratch length can be up to 20 mm with a lateral force specification to 1 N with resolution of 30 μN.

11.6 "UMIS®," CSIRO[15]

The UMIS® is a load-controlled nanoindentation instrument. A particular feature of the instrument is the patented[16] parallel spring load feedback system that ensures that the applied load is held equal to the set load. This separation of the load measurement and actuation ensures that the actual load applied to the indenter shaft is recorded. A photograph of the unit with the isolation stand and environmental enclosure removed is shown in Fig. 11.7. The heavy construction of the UMIS® acts as a seismic mass for the damped support springs that isolate the instrument from mechanical vibration and minimize compliance of the load frame. The environmental enclosure is insulated with glass wool and an electromagnetic screen to minimize thermal and electrical interference.

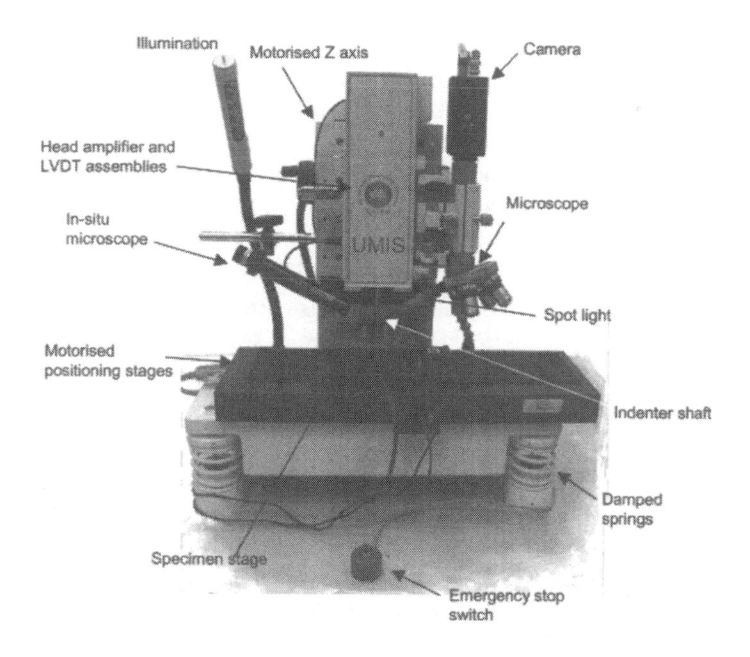

Fig. 11.7 The "UMIS®" (Courtesy CSIRO).

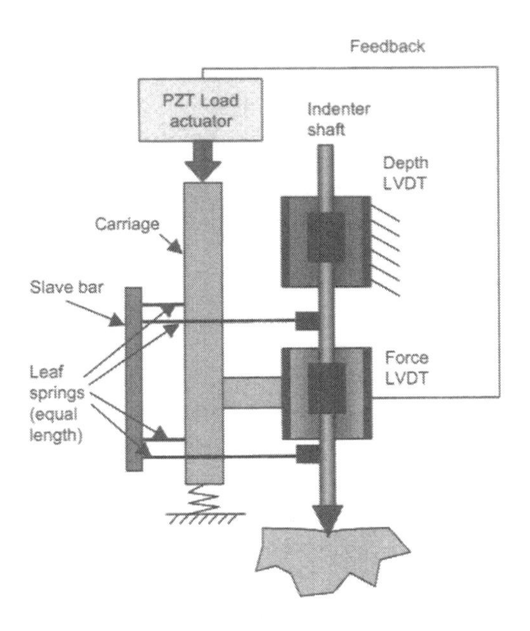

Fig. 11.8 Schematic of the operating principle of the UMIS®. An electronic feedback system measures the force from the force LVDT and adjusts the expansion of the PZT until the requested or commanded force is actually achieved at the indenter shaft.

The UMIS® applies load via the expansion of a piezoelectric element connected to the indenter shaft by a series of carefully machined leaf springs. The deflection of the springs is a measure of the load applied to the indenter and this deflection, along with the displacement of the shaft relative to the loading frame, is done using very high-quality linear variable differential transformers (LVDTs). Figure 11.8 shows a schematic of the operating principle used in the UMIS®. Since the depth LVDT measures the actual indenter displacement, the compliance of the load cell is not recorded as part of the overall displacement signal as in conventional load cell arrangements. The over-damped force-feedback loop between the load cell and the load actuator eliminates any non-linearities associated with the PZT element.

As with other instruments of this type, an optical microscope is fitted to allow for precise specimen positioning and post-indentation crack length measurement. The instrument is software controlled and can be run in an automated, unattended mode for multiple indentations on multiple specimens. Specimen positioning is by means of servo-controlled dc motor driven positioning stages. The same stage allows the specimen to be translated from the indenter position to the optical microscope and alternately to the optional AFM attachment. Several brands of AFM can be attached to the UMIS®.

The WinUMIS control software uses the partial-unload technique for measuring the contact stiffness at each load increment that provides both elastic modulus and hardness of the specimen as a function of depth of penetration. Provision is made for thermal drift correction, initial penetration, compliance determination, and indenter area function calibration. The software also provides a simulation mode of operation whereby mechanical properties of the specimen and indenter are used as inputs to calculate a theoretical load-displacement response that may be compared with experimental results. The control software allows any combination of loading and unloading to take place through the use of "script" files that can be assembled automatically for standard tests or created by the user as desired. The UMIS® is a dual range instrument with ×1 and ×10 on load and force ranges selectable by the user.

The specifications of the instrument are as follows:

Maximum load	50 mN (Range A), 500 mN (Range B)
Minimum contact force	2 μN
Force resolution	500 nN
Force noise floor	750 nN
Maximum depth	2 μm (Range A), 20 μm (Range B)
Depth resolution	0.03 nm
Depth noise floor	0.05 nm
Specimen positioning	± 0.5 μm (±0.1 μm optional)
Field of testing	50×50 mm (optional rotational vacuum table for up to 12" wafers).
Load frame compliance	0.1 nm/mN
User interface	Windows® 95/98/NT/2000

The UMIS® can be configured for additional ranges from 1 mN to a maximum of 5 N and depths of up to 40 μm or more.

The UMIS® is offered with optional components such as a scratch tester and a 500 °C hot stage. In the scratch tester, lateral force is measured with another separate LVDT displacement transducer via a series of calibrated leaf springs. Friction coefficient can be measured as a function of scratch length in either a ramped or steady load traverse. The force feedback operation of the UMIS® allows the instrument to operate as a surface profilometer. In high temperature testing, temperature of the sample is controlled with a closed loop temperature controller. The WinUMIS software allows the indenter tip to be held in contact with the specimen surface at the initial contact force to ensure thermal equilibrium of the contacting parts. The electronic force feedback control maintains the initial contact force regardless of changes in dimensions of the parts during this procedure. The main specimen heater has a water-cooled exterior to maximize temperature stability of the system and to protect the positioning stages from excessive temperature.

The unique features of the UMIS® that distinguish it from its competitors are the exceptional repeatability, stability, and robustness of the complete system. This makes the unit extremely easy to use, even by inexperienced operators. The robustness of the design makes the UMIS® suitable for use in industrial, research, and teaching laboratories.

References

1. "MTS," "Nano," and "Nano Indenter" are registered trademarks of MTS Systems Corporation, Nano Instruments Innovation Center, 1001 Larson Drive, Oak Ridge, TN 37830 USA.
2. J.B. Pethica, "Microhardness tests with penetration depths less than ion implanted layer thickness in ion implantation into metals," Third International Conference on Modification of Surface Properties of Metals by Ion-Implantation, Manchester, England, 23-26, 1981, V. Ashworth et al. eds. Pergammon Press, Oxford, 1982, pp. 147–157.
3. B.N. Lucas, W.C. Oliver, and J.E. Swindeman, "The dynamics of frequency-specific, depth sensing indentation testing," Mat. Res. Soc. Symp. Proc. 522, 1998, pp. 3–14.
4. W.C. Oliver and J.B. Pethica, "Method for continuous determination of the elastic stiffness of contact between two bodies," United States Patent, 4848141, 1989.
5. COMOS is a registered trademark of Structural Research and Analysis Corporation.
6. Micro Materials Ltd., Unit 3, The Byre, Wrexham Technology Park, Wrexham LL13 7YP, United Kingdom.
7. Hysitron Inc., 10025 Valley View Road, Minneapolis, MN, 55344 USA.
8. W.A. Bonin and Hysitron Inc., "Apparatus for microindentation hardness testing and surface imaging incorporating a multi-plate capacitor system," United States Patent, 5553486, 1996.
9. W.A. Bonin and Hysitron Inc., "Capacitive transducer with electrostatic actuation," United States Patent, 5576483, 1996.
10. W.A. Bonin and Hysitron Inc., "Multi-dimensional capacitive transducer," United States Patent, 5661235, 1997.
11. W.A. Bonin and Hysitron Inc., "Multi-dimensional capacitive transducer," United States Patent, 5869751, 1999.
12. W.A. Bonin and Hysitron Inc., "Apparatus for microindentation hardness testing and surface imaging incorporating a multi-plate capacitor system," United States Patent, 6026677, 2000.
13. CSM Instruments, Rue de la Gare 4, CH-2034 Peseux, Switzerland.
14. N. X. Randall, C. Julia-Schmutz, J. M. Soro, J. von Stebut, and G. Zacharie, "Novel nanoindentation method for characterising multiphase materials," Thin Solid Films, 308–309, 1997, pp. 297–303.
15. CSIRO, Bradfield Rd, West Lindfield, NSW 2070 Australia.
16. J.S. Field, "Penetrating measuring instrument," United States Patent, 5067346, 1991.

Chapter 12
Applications of Nanoindentation

12.1 Introduction

Nanoindentation finds a wide application. The test results provide information on the elastic modulus, hardness, strain-hardening, cracking, phase transformations, creep, fracture toughness, and energy absorption. Since the scale of deformation is very small, the technique is applicable to thin surface films and surface modified layers. In many cases, the microstructural features of a thin film or coating differs markedly from that of the bulk material due to the presence of residual stresses, preferred orientations of crystallographic planes, and the morphology of the microstructure. The proceedings of annual symposiums are a rich source of information about the applications of nanoindentation. In this chapter, some rather straightforward examples of analysis of nanoindentation test data are presented using the methods described in the previous chapters.

The two most commonly measured properties during an nanoindentation test are elastic modulus and hardness. Both may be measured as a function of depth of penetration into the specimen surface, thus providing a depth profile of these properties. Hardness is important since it is related in many cases to the strength or fracture toughness of the specimen. A high hardness generally corresponds to a high abrasive wear resistance. Hardness and modulus values can also be used to monitor surface or material consistency. Elastic modulus provides a measure of stiffness or compliance of the specimen. Hard materials are usually also very brittle and a hard protective coating may often be deposited on a relatively soft material with a high elastic modulus to provide a rigid support that will help prevent brittle fracture. Measurements of modulus of both film and substrate allow such thin film systems to be optimized for a particular application. The ratio of modulus and hardness (E/H) also provides valuable information about a material since it is this ratio that determines the spatial extent of the elastic deformation that might occur under loading before permanent yielding occurs.

The extremely high sensitivity of the load and displacement sensors in a typical nanoindentation instrument makes such devices suitable for MEMs testing in some cases. However, for this application, it should be recognized that the instrument circuits are usually optimized for contacts on relatively stiff specimens. For example, feedback control parameters may not be optimized for compliant structures such as cantilevers in a MEMs device.

12.2 Fused Silica

Perhaps one of the most common and easiest specimens to test using nanoindentation is optical grade fused silica‡‡‡. Polished specimens are readily available as discs that can be easily mounted and tested. A maximum load of 50 mN with 25 load increments will result in a depth of penetration of about 650 nm for a Berkovich indenter, well within the capability of any good nanoindentation instrument.

Figure 12.1 shows the load-displacement curve obtained with a Berkovich indenter on a fused silica specimen. In this case, for a load of 50 mN, the uncorrected depth of penetration was found to be 629.9 nm. After correction for instrument compliance, the penetration was 624.9 nm. An initial penetration depth of $h_i = 6.5$ nm was determined using the logarithmic fitting method (see Chapter 4) for an initial contact force of $P_i = 10$ μN leading to a maximum depth of penetration for this test of $h_t = 631.4$ nm.

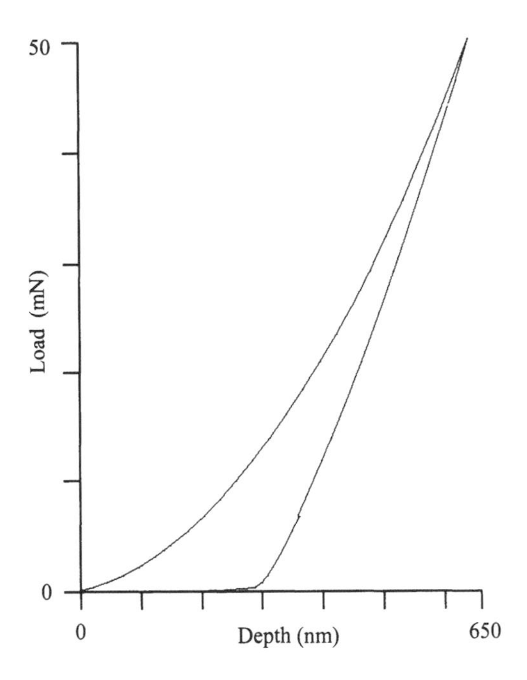

Fig. 12.1 Load-displacement curves for a nanoindentation experiment on fused silica for Berkovich indenter at 50 mN maximum load. Data have been corrected for thermal drift, load frame compliance, and initial penetration. Maximum depth of penetration is 627 nm (Courtesy CSIRO).

‡‡‡ Fused silica is amorphous SiO_2 manufactured by chemical combination of silicon and oxygen. Fused quartz is a similar material made from melting natural silica crystals or sand and is considered to be less pure than fused silica.

Using the first eight unloading points for the fitting of the data for the initial unloading slope, we find a value for dP/dh of 185.9 mN/μm with an estimated residual depth, h_r, of 59.0 nm and a plastic depth h_p equal to 423.0 nm. Using the multiple point unload method, this leads to a value for the combined modulus of the system E^* of 69.68 GPa from which follows an estimation of the specimen modulus of 72.45 GPa using a modulus of 1010 GPa and a Poisson's ratio of 0.07 for the diamond indenter. The hardness is computed to be 9.54 GPa. These results are in reasonable agreement with the accepted values of E = 72.5 GPa and H = 7.6 GPa (the latter obtained by optical methods). In depth-sensing indentation work, H = 9 – 9.5 GPa is usually obtained for this material.

While many practitioners use fused silica as a reference material for calibration and verification of nanoindentation instruments and indenters, it is not often appreciated that the material exhibits time-dependent creep under indentation loading. For tests with a cube corner indenter, at a constant load of 10 mN, a change in depth in the order of 5 nm over 10 seconds can be expected. Reference testing with fused silica should include a hold period of at least 10 to 15 seconds at maximum load before unloading so as to minimize the effect of creep on the slope of the unloading data.

It should be appreciated that this test, trivial as it may seem, provides a very good example of induced plasticity in what is nominally a brittle material.[1] The existence of a residual impression is proof of plastic deformation within the material.

12.3 Titanium Dioxide Thin Film

A common application of nanoindentation testing is the measurement of properties of thin film specimens. This is not often an easy task, since the films of interest are often on the order of 500 nm thickness or less. Using the 10% rule, the maximum depth of penetration is usually limited to about 50 nm or less for hardness determinations. In the work reported here, 500 nm thick films of TiO_2 were deposited on silicon substrates using a filtered arc deposition process in which the process parameters were changed in order to study the resulting change in mechanical and optical properties. Using a gradually increasing bias voltage, Bendavid, Martin, and Takikawa[2] found that an observed change in hardness and modulus of the films corresponded to a change in phase of the film material from anatase to rutile forms. By increasing the bias voltage, the energy of the bombarding ions in the arc deposition process is increased, and this causes structural modifications to the microstructure that increase its hardness.

Figure 12.2 shows how the measured film hardness (for nanoindentation tests done at 2 mN maximum load) resulted in a change of hardness from 11.6 GPa to 17.6 GPa over a range of bias voltages from zero to –400 V. Testing of this kind enables the mechanical properties of such films to be tailored to particular applications.

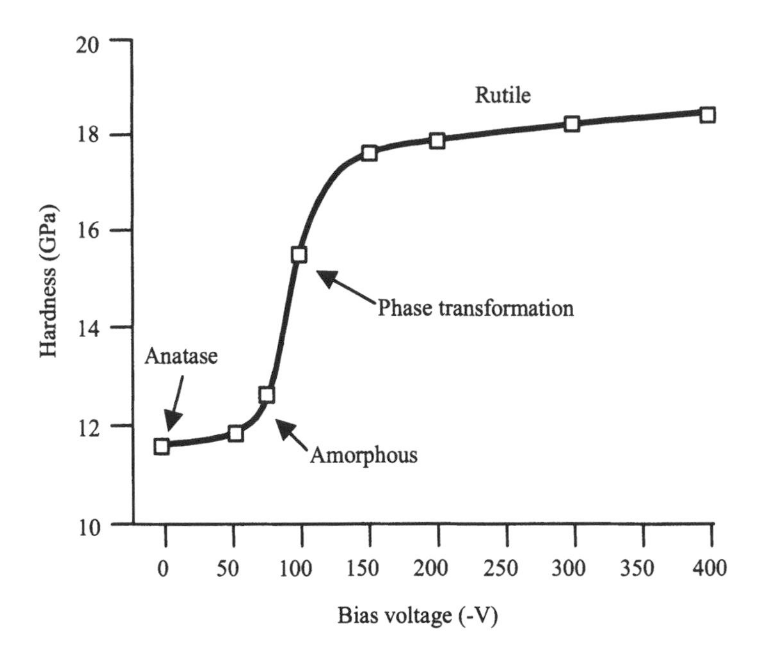

Fig. 12.2 Hardness of TiO_2 films on a Si substrate as a function of substrate bias voltage. A rapid increase in hardness is observed at about –100 V bias which coincides with a phase transformation from anatase to rutile (Courtesy CSIRO and after reference 2).

12.4 Superhard Thin Film

An emerging application of nanoindentation with respect to thin films is the hardness testing of very hard coatings. The hardness of hardened steel is in the order of 10 GPa, and of ceramics is on the order of 20 – 30 GPa. Recent developments in thin film technology have resulted in measured indentation hardness values of greater than 40 GPa. Such materials are called "superhard." The hardness of nanostructured materials have been measured at greater than 50 GPa and Veprek[3] reports a measured hardness of a nc-TiN/SiN material of 105 GPa, harder than that of diamond (usually taken to be 100 GPa).

Figure 12.3 shows the load displacement curves for two thin film systems, the first, a conventional TiN film, and the second, a nanocomposite film. In the nanocomposite material, the amorphous silicon phase serves to restrict the growth of TiN structures, and this has the effect of increasing the hardness of the composite material. The measured hardness in this case for the nanocomposite film is approximately 60 GPa. Note that for a maximum load of about 4 mN, the penetration depth for these types of coatings is about 80 nm.

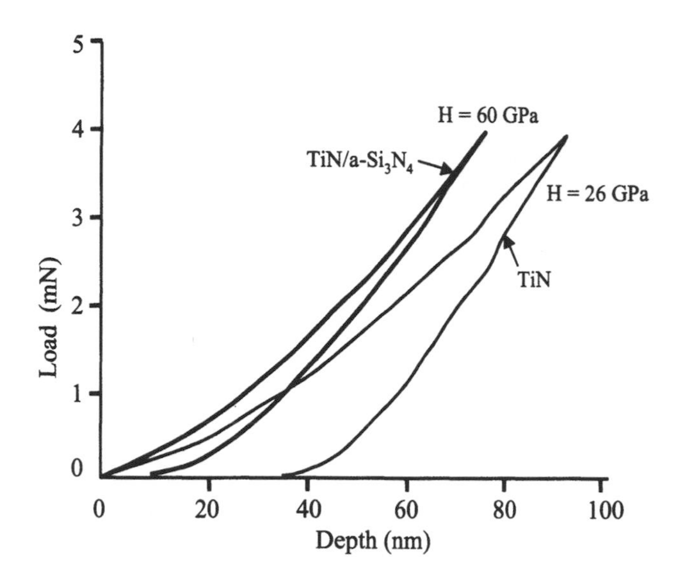

Fig. 12.3 Load-displacement response of TiN thin films, one doped with an amorphous silicon phase that serves to restrict the growth of the TiN grains during the deposition process and this provides an effective increase in hardness (Courtesy CSIRO).

The use of nanoindentation for the measurement of superhard materials is a particularly difficult challenge. The depths of penetration are on the order of the radius of the tip of a typical Berkovich indenter and are also comparable to the surface roughness of the specimen surface. Such thin films are often prepared with a considerable level of residual stress that also affects the indentation hardness value. Attention to the methods of correction and factors affecting nanoindentation test results described in Chapters 3 and 4 have to be taken seriously indeed if values of hardness in excess of 40 GPa are to be considered meaningful rather than just comparative. An interesting theoretical and practical challenge is how to measure the hardness of materials that may well be harder than diamond – traditionally considered the hardest known substance.

12.5 Diamond-like Carbon (DLC) Thin Film

Diamond-like carbon is a very important material for anti-wear films in the semiconductor and computer industry. Film thicknesses are typically on the order of 10 nm. In these cases, substrate effects are almost impossible to avoid during indentation testing. Scratch testing of DLC films is a useful alternative to evaluate adhesion and mechanical properties.

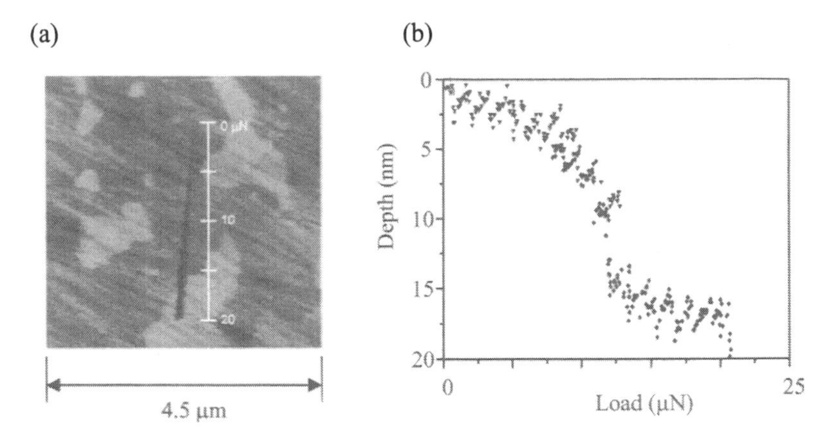

Fig. 12.4 (a) Residual impression from a scratch test in a 7.5 nm DLC film using a ramped force up to 20 μN. (b) Penetration depths for multiple scratch tests plotted against normal load F_N. Failure of film is evidenced by sudden increase in depth at a load of about 12 μN (Courtesy Hysitron Inc.).

Figure 12.4 (a) shows a ramped load scratch from 0 up to 20 μN on a 7.5 nm DLC film on a computer hard disk head slider. Figure 12.4 (b) shows the normal displacement as a function of the normal force. At a critical load of about 12 μN, the film fails and a sudden increase in penetration depth is recorded.

12.6 Creep in Polymer Film

One of the less well-known applications of nanoindentation is the estimation of the viscoelastic properties of biomaterials and polymers. Figure 12.5 shows a conventional load-displacement curve obtained with a spherical indenter on a polymer film. In this case, creep within the specimen material leads to a negative slope for the unloading, and this invalidates the conventional methods of analysis given in Chapter 3.

In Chapter 7, it was shown how a simple three-element model could be used to extract viscoelastic properties of the specimen from the creep curve of a nanoindentation test. A creep curve is obtained by holding the load constant (usually at maximum load) and monitoring the change in depth. Figure 12.6 shows creep curves obtained on two specimens of a copolymer, the second specimen having a cross-linker added. In these tests, a 20 μm sphero-conical indenter was loaded to 1 mN and held for 10 seconds. The theoretical creep curves shown in Fig. 12.6 were generated using Eqs. 7.3d, 7.3f and 7.3h in conjunction with the non-linear least squares fitting procedure described in Appendix 4.

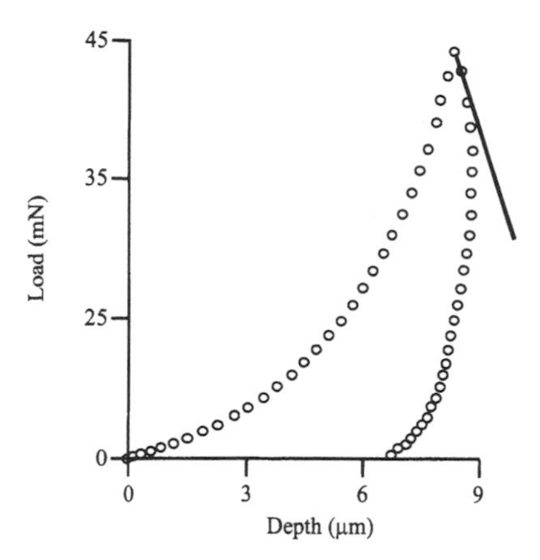

Fig. 12.5 Conventional load-displacement curve resulting from indentation of a polymer film with a 20 μm radius spherical indenter. Creep within the specimen material during the test results in a negative slope of the unloading curve.

The results in Table 12.1 indicate that the cross-linked resin is stiffer and more solid-like than the base copolymer, but both materials exhibit creep behavior during an indentation test. The conventional method of analysis using the slope of the unloading curve would not provide meaningful information about the mechanical properties of these materials.

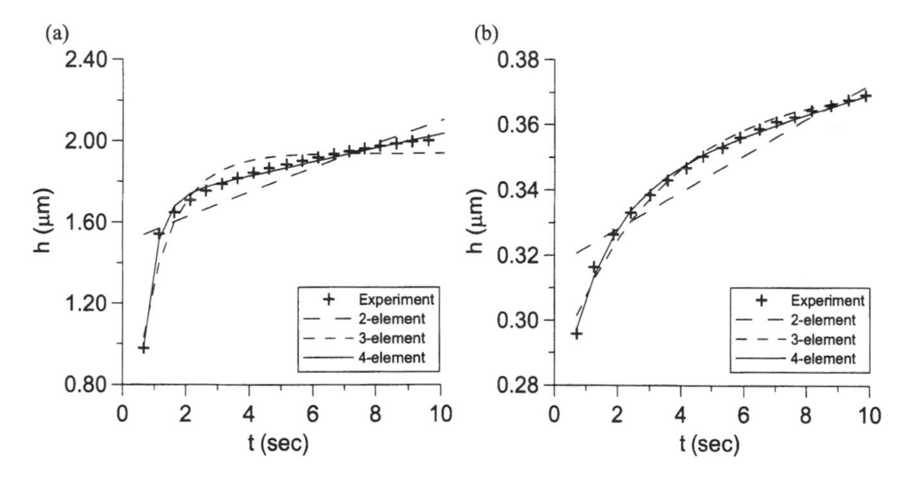

Fig. 12.6 Creep curves for (a) base copolymer and (b) cross-linked material using a 20 μm sphero-conical indenter with an applied load of 1 mN over 10 seconds. Data points indicate experimental results, solid lines indicate best fit to Eq. 7.3d, 7.3f and 7.3h (Courtesy CSIRO).

Table 12.1 Results of creep analysis for base copolymer and cross-linked material. Parameters E and η were found using the least squares fitting procedure described in Appendix 4.

Model:	2-element		3-element			4-element			
GPa/GPas	E_1	η	E_1	E_2	η	E_1	E_2	η_1	η_2
Base	0.09	1.69	0.16	0.103	0.135	0.17	0.14	2.9	0.7
Cross-linked	0.94	41.2	1.03	2.73	11.2	1.05	4.7	72.6	7.7

12.7 Fracture and Delamination of a Silicon Oxide Film

Nanoindentation testing can be used not only to measure hardness and modulus of thin films, but also to induce a controlled amount of damage within the film and/or the substrate. As mentioned in Section 2.4, an experienced person can relate features on the load-displacement curve to particular modes of damage in the specimen.

In the example shown in Fig. 12.7, a 1 μm sphero-conical indenter has been used to intentionally fracture a 1 μm thick SiO_2 film on silicon. Delamination is represented by the light area surrounding the indentation, and radial cracks are evident. These events manifest themselves as discontinuities on the load-displacement curve.

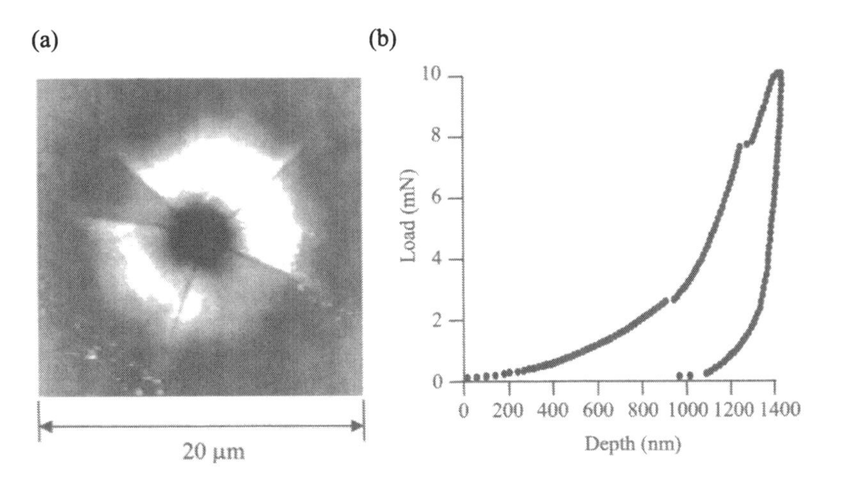

Fig. 12.7 (a) The result of an indentation with a 1 μm sphero-conical tip into 1 μm SiO_2 film on silicon. (b) Pop-in events on a load-displacement curve can be used to quantify the conditions for the initiation of cracks and delamination (Courtesy Hysitron Inc.).

12.8 High-Temperature Testing on Fused Silica

In most materials, the elastic modulus decreases with increasing temperature. One unusual property of fused silica is that its elastic modulus increases with increasing temperature. Figure 12.8 (a) shows a load-displacement response of a soda-lime glass specimen at room temperature and at 200 °C. Figure 12.8 (b) shows the variation in modulus with temperature for fused silica where the temperature ranged from room temperature to 400 °C. Henley, Mao, Bell, and Mysen[4] showed that under hydrostatic pressures of up to 8 GPa a reversible structural change occurred corresponding to a change in the Si-O-Si bond angle. Above 8 GPa, the structural change was irreversible. Galeener[5] showed that a change in the bond angle also occurred at temperatures above 900 °C. Thus, in an indentation test on fused silica, both hydrostatic pressure and temperature play a role in this behavior.

There are many other examples which justify high temperature measurements, including the investigation of microelectronic thin films normally processed at temperatures up to 500 °C, indentation creep studies, scratch testing of polymer coatings and other thin films, and the investigation of temperature-sensitive fibre-matrix bonding forces. In addition, dislocation mobility (and therefore hardness) is temperature sensitive.

From a practical standpoint, the wear performance of a film-substrate combination will ideally require optimization of the relative properties of the film and substrate materials at the service temperature rather than inferring their properties from room temperature results.

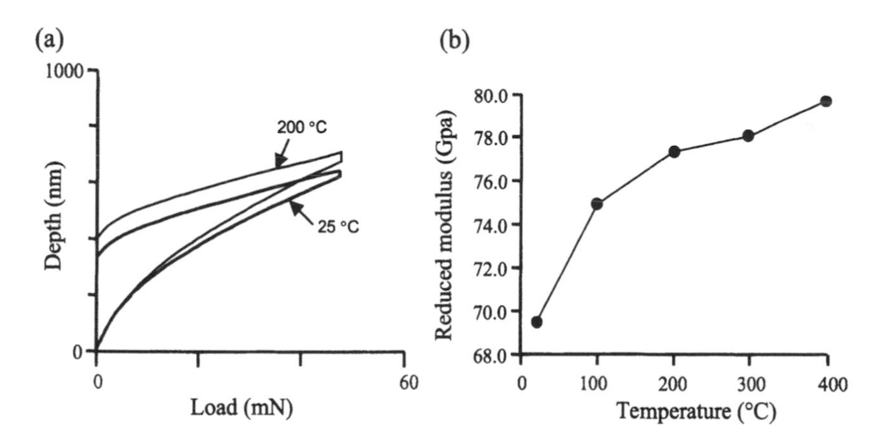

Fig. 12.8 (a) Load-displacement curve for soda-lime glass at 200 °C. (b) Variation in elastic modulus computed from nanoindentation test results on fused silica as a function of temperature (Courtesy Micro Materials Ltd.).

12.9 Adhesion Measurement

A particularly novel application of nanoindentation testing is the measurement of adhesion for powder particles. This application is of particular relevance in the pharmaceutical industry. Powder-coated specimens in the form of cylinders (e.g., wires coated with the powder) are attached to the specimen stage and indenter mounting. The cylinders are then bought into contact with a very low force. The force is then slowly reduced through zero to negative (pulling) values until the surfaces separate. The force at separation is identified and is a measure of the adhesion of the particles due to van der Waals attractions (see Appendix 2). The use of coated surfaces together ensures a statistically more meaningful result than a measurement with single particles.

A typical adhesion result using the NanoTest® instrument is shown in Fig. 12.9. Here, the depth and load were monitored continuously as the load was increased, held, and then decreased through zero to negative values until eventually the pendulum moved away from the specimen surface. The maximum applied load was 250 μN prior to load reversal for specimen separation. The adhesion force was found to range from 0.14 mN to 0.19 mN for tests on different parts of the specimen.

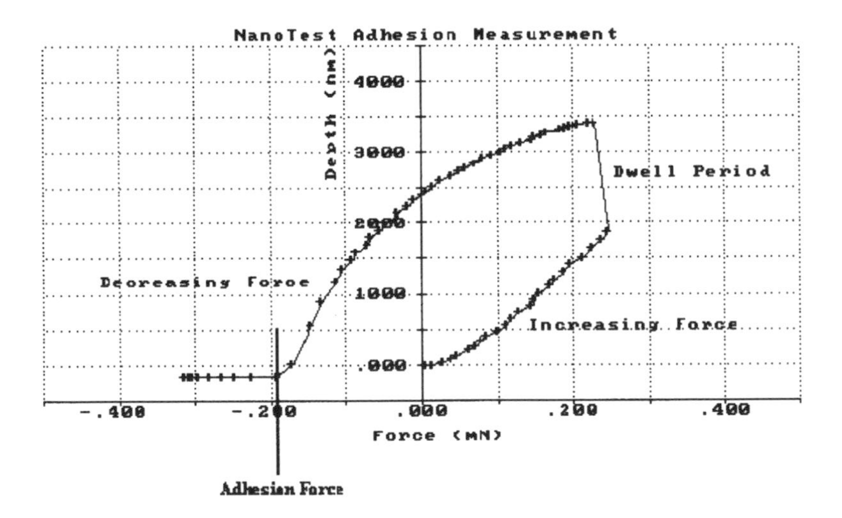

Fig. 12.9 Adhesive force result using the NanoTest® pendulum arrangement. The load was increased, held, and then decreased through zero to negative values until the surfaces separated (Courtesy Micro Materials Ltd.).

12.10 Dynamic Hardness

The "dynamic hardness" is defined as the energy consumed during a rapid indentation divided by the volume of the indentation. The energy of the indentation can be determined from the ratio of the impact and rebound velocities as the indenter bounces on the specimen surface. The volume of indentation is found from the penetration depth and the known geometry of the indenter.

Figure 12.10 shows the results for an aluminium specimen with a Berkovich indenter. The impact and rebound velocities are obtained from the slopes of the responses just before, and just after, impact. For the specimen tested, the quasi-static hardness was measured to be 0.27 GPa while the dynamic hardness was measured to be 0.54 GPa. This type of result is relevant for dynamic processes such as impact wire bonding.

Successful dynamic hardness testing requires a relatively low system resonant frequency.

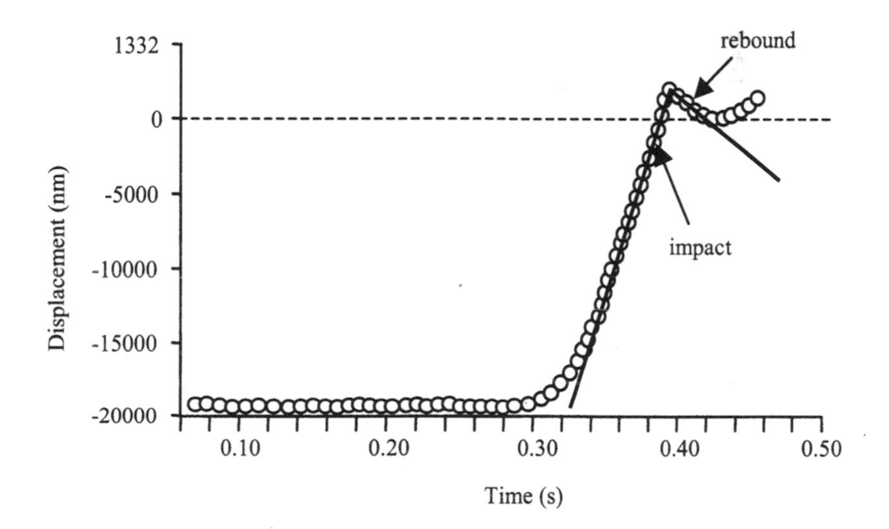

Fig. 12.10 Displacement vs time measurements during impact and rebound for an aluminium specimen with a Berkovich indenter (Courtesy Micro Materials Ltd.).

12.11 Repeatability Testing

It is of interest to purchasers of nanoindentation instruments to be able to evaluate their relative features and performance. A very good test of repeatability of an instrument is the comparison of load-displacement curves for different maximum loads. Figure 12.11 shows such a comparison.

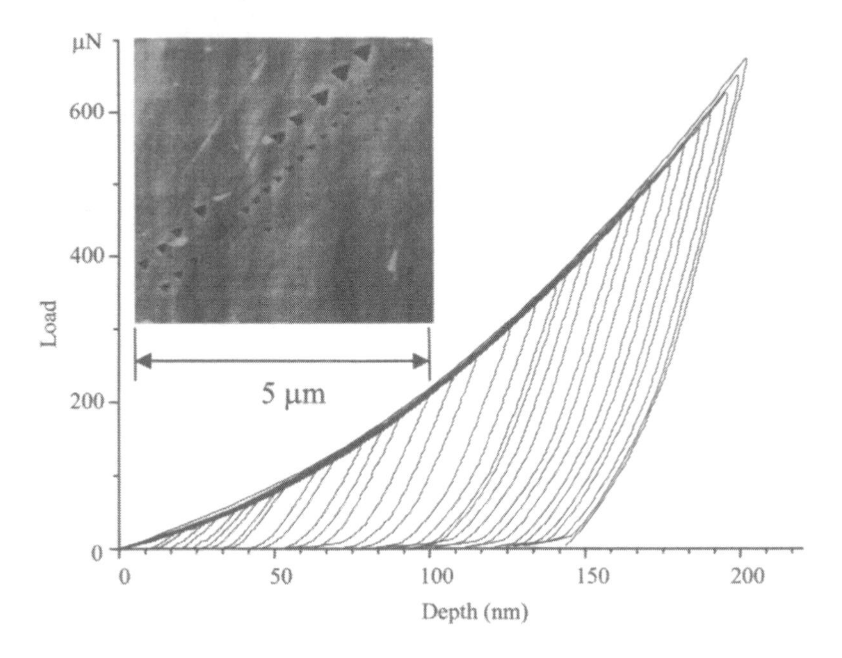

Fig. 12.11 Load-displacement curves for varying maximum loads on fused silica. The curves should all fall on top of one another for the loading half of the cycle and show a smooth and regular spacing for the unloading portion of the response. The inset shows the residual impressions made in the surface for the load-displacement curves shown (Courtesy Hysitron Inc.).

An overlay of the load-displacement curves for a wide range of loads should all fall within acceptable limits on the loading curve, and show a smooth response for both the loading and unloading portions. The inset in Fig. 12.11 shows the residual impressions in the specimen surface for the load-displacement curves shown.

12.12 Assessment of Thin Film Adhesion by Scratch Testing

The need for quantitative assessment of the adhesion of thin films and coatings is a pressing one as manufacturers of such structures are called upon to increase the film performance while decreasing their thickness. Conventional scratch tests involve the measurement of the critical load for failure under an increasing load. The critical load can be determined by inspection or, in some cases, by a sudden change in the friction coefficient, acoustic emission, or penetration depth.

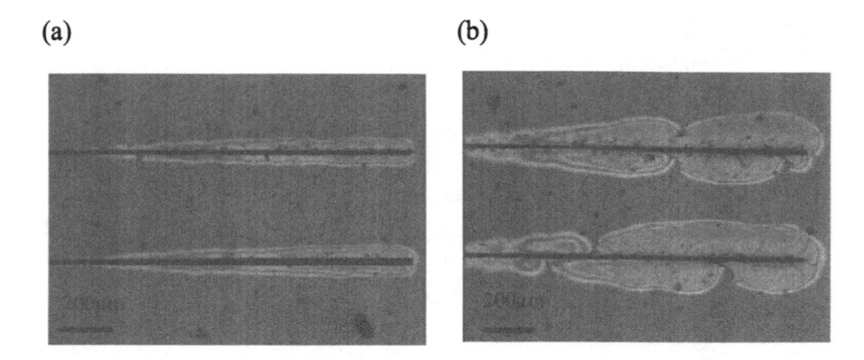

Fig. 12.12 A comparison of two TiO_2 coated glass substrates subjected to a progressive normal load scratch of 0-50 mN with a 10 μm diamond indenter. The scratch speed is 2.5 mm/minute at a length of 2.5 mm. (a) High roughness specimen; (b) low roughness specimen (Courtesy CSM Instruments).

Figure 12.12 shows the results of scratch tests involving a ramped, increasing force applied to a 10 um spherical indenter on two titanium oxide thin films. The scratch length is about 2.5 mm while the normal force varied from zero to 50 mN. Conventional large-scale scratch testers are limited by the magnitude of the forces at the low end of the range, while AFM-based instruments are often limited by their short range of scratch length.

12.13 AFM Imaging

Very useful materials information can be obtained from AFM imaging of the residual impression in the specimen surface. In Fig. 12.13, residual impressions for fused silica, hardened steel and sapphire are shown. Note the relative size of the impressions that have all been made at the same load. Piling-up can be seen in the steel specimen. Sinking-in is evident in the fused silica sample.

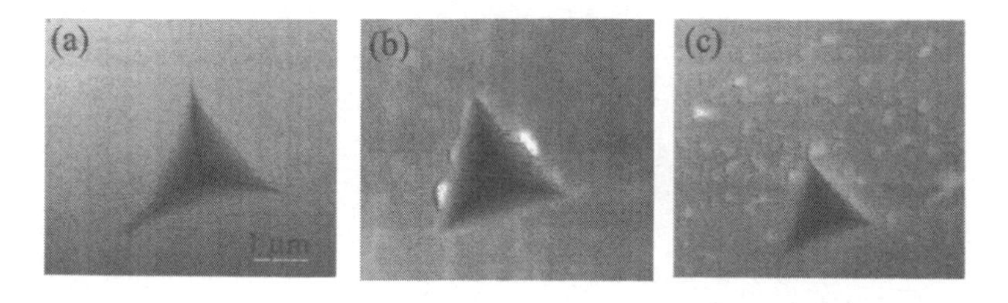

Fig. 12.13 AFM scans made with a DME Dualscope® AFM objective of residual impression at 50 mN with a Berkovich indenter in (a) fused silica, (b) steel and (c) sapphire (Courtesy CSIRO).

12.14 Fracture Toughness

Measurement of fracture toughness in very small volumes of materials is made possible using nanoindentation because of the availability of very sharp, and very accurately dimensioned, indenter tips. The cube corner indenter is particularly useful for fracture toughness measurements. Nanoindentations were carried out using a Hysitron TriboScope® with a NorthStar® cube corner diamond indenter tip under varying loads. It was found that a threshold load of 10 mN existed below which cracks were not observed. Fig. 12.14 shows one of residual impressions using the in-situ imaging technique for a load above the threshold load. The relevant features of the image are the cracks emanating from the corners of the indentation. In this example, the fracture toughness was evaluated using:

$$K_c = k\left(\frac{E}{H}\right)^n \frac{P}{c^{3/2}} \tag{12.14a}$$

with the empirical constant k being set to 0.032. The force-displacement curves for the indentations without cracks (below the threshold load) were used to get the hardness H and elastic modulus E for use in the formula. The in-situ post-indentation images were obtained immediately after indentation. Then, the lengths of the cracks, if any, in the indents, were measured directly from the in-situ images.

The fracture toughness was found to be equal to 3 MPa $m^{1/2}$ and relatively independent of load up to 40 mN once a threshold load of 10 mN to induce cracking was achieved.

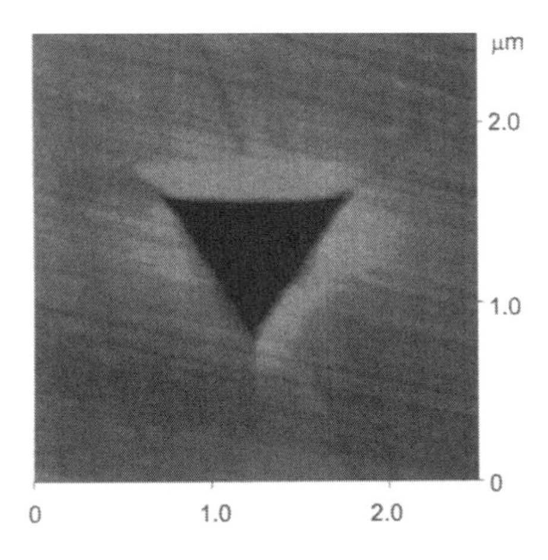

Fig. 12.14 In-situ image of indentation in SiC with a NorthStar® cube-corner indenter. Image and indentation made with a Hysitron Triboscope® (Courtesy Hysitron Inc.)

12.15 Thermal Barrier Coating

A particularly important reason for performing nanoindentation tests at high temperature is to determine the mechanical properties of the sample at the temperature are which they are likely to operate.

Fig. 12.15 shows load-displacement curves from indentations at 750 °C in a thermal barrier coating comprising zirconia and yttria. A notable feature of these curves is the relatively smooth nature of pop-in events in the loading curve. At high temperatures, more of these events were observed compared to the same test conducted at room temperature. Normally, pop-in events are characterized by a sudden increase in displacement at a particular load, whereas here, there is a more gradual increase in depth over a relatively small load increment. It is conjectured that this behaviour represents slow crack growth during the indentation loading.

Measurements of hardness and modulus over the surface of the specimen yielded results in the range 2.2 – 3.8 GPa and 75 – 140 GPa respectively. When ranked by the magnitude of E and H, plateaux in values within these ranges were observed indicating the presence of different phases within the material.

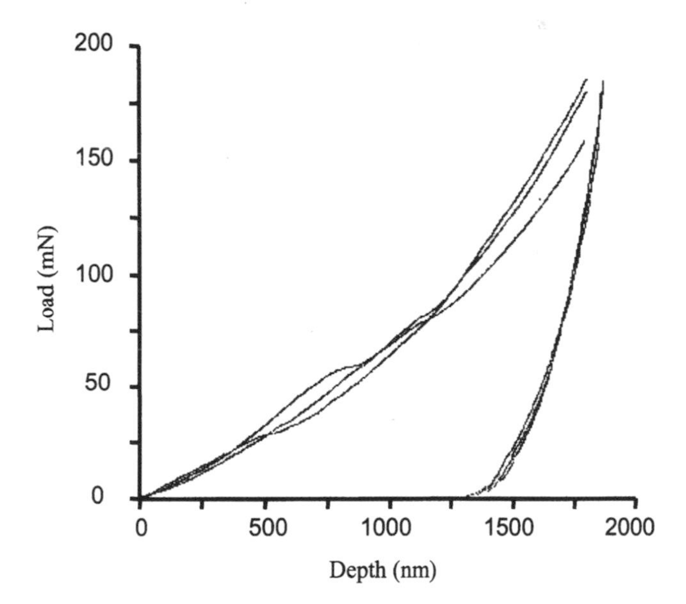

Fig. 12.15 Load-displacement curve for zirconia – yttria thermal barrier coating held at 750 °C. Smooth pop-in events on the loading curves are believed to be an indication of slow crack growth. (Courtesy Micro Materials Ltd.)

12.16 Pile-up in Aluminium Coating on Silicon

Fig. 12.16 shows SFM images of nanoindentation with a Berkovich indenter on an aluminium coating deposited on silicon taken with a Nano-Hardness Tester®. The SFM images show that very little pile-up is evident at low depths, this being characteristic of sputtered aluminium. The evolution of cross-sectional profiles shows a variation in residual depth for the silicon substrate below the surface. At the greatest penetration depths, the observed transition between the film and the substrate is nearer to the surface owing to the increased elastic relaxation of the latter upon unloading. This is an important factor that should always be considered when estimating the thickness of a coating from the profile of a residual indentation, especially for the case of substrate materials which exhibit large elastic recovery. For the system tested here, an estimate of the film thickness from the deepest profile would suggest a value of approximately 220 nm where the thickness from the deposition rate is known to be 300 nm.

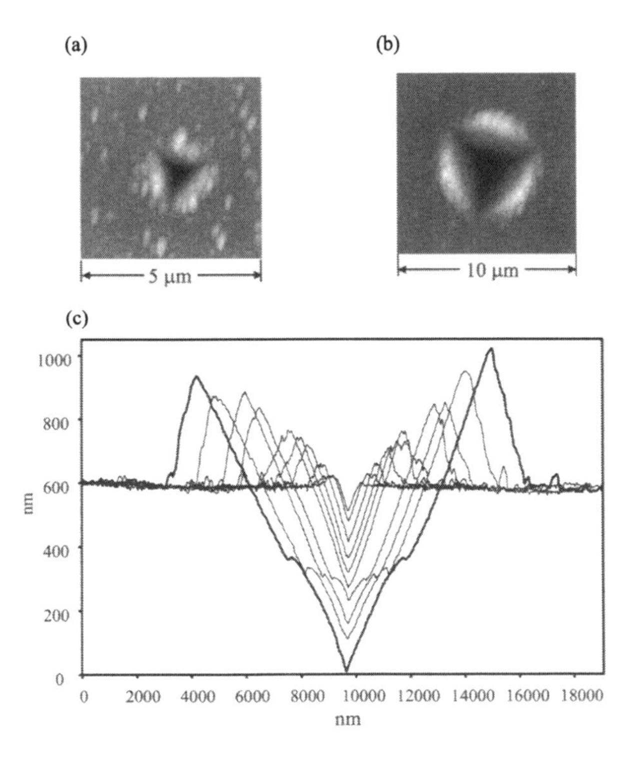

Fig. 12.16 Examples of SFM imaging showing pile-up of a 300 nm aluminium film on a silicon substrate. (a) Surface impression at penetration depth 100 nm (b) Surface impression at penetration depth 600 nm. (c) Evolution of surface profile of the impression at a range of penetration depths[6]. (Courtesy N.X. Randall, CSM Instruments.)

12.17 Other Applications

Figure 12.17 shows some further fine examples of nanoindentation testing.

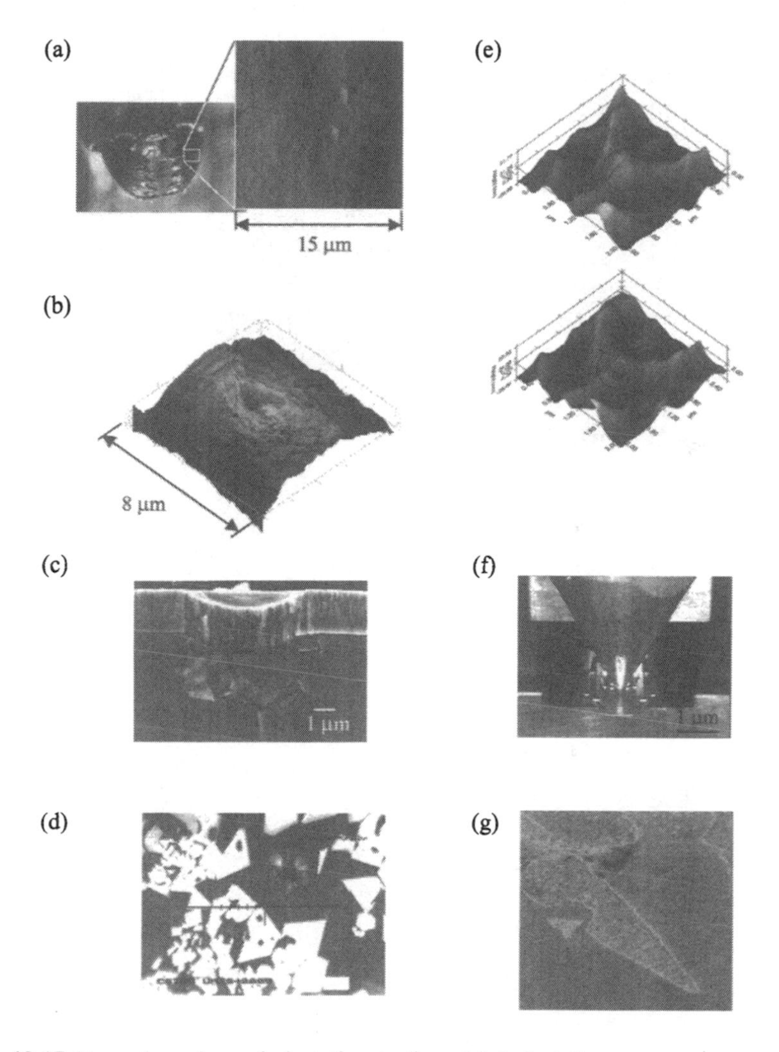

Fig. 12.17 Examples of nanoindentation testing. (a) Indentations on an ant mandible similar to those shown in left image. (Courtesy Hysitron Inc.) (b) Optical fiber. (Courtesy Hysitron Inc.) (c) Cross-section of controlled damage in 2.7 μm TiN film on silicon using 5 μm spherical indenter at 350 mN load. (Courtesy CSIRO and after reference 7) (d) Precise positioning of indentation on a multiphase iron ore sinter and measurement of fracture toughness. (Courtesy CSIRO), (e) In-situ SPM images of collapse of a Cu nano-tunnel on TaS_2 substrate under indentation loading. (Courtesy Hysitron Inc.) (f) Flexure testing of hard disk read/write head. (Courtesy CSIRO) (g) Modulus map of artificial tooth material. (Courtesy Hysitron Inc.)

References

1. D.M. Marsh, "Plastic flow in glass," Proc. Roy. Soc. A279, 1964, pp. 420–435.
2. A. Bendavid, P.J. Martin, and H. Takikawa, "Deposition and modification of titanium dioxide thin films by filtered arc deposition," Thin Solid Films, 360, 2000, pp. 241–249.
3. S. Veprek "The search for novel, superhard materials," J. Vac. Sci. Technol. A17 5, 1999, pp. 2401–2420.
4. R.J. Hemley, H.K. Mao, P.M. Bell, and B.O. Mysen, "Raman-Spectroscopy of SiO_2 glass at high pressure," Phys. Rev. Lett. 57 6, 1986, pp. 747–750.
5. F.L. Galeener, "Raman and ESR studies of the thermal history of amorphous SiO_2," J. Non-Cryst. Solids, 71, 1985, pp. 373–386.
6. N.X. Randall, "Direct measurement of residual contact area and volume during the nanoindentation of coated materials as an alternative method of calculating hardness", Phil. Mag. A, 82 10, 2002, pp. 1883–1892.
7. E. Weppelmann, M. Whittling, M.V. Swain, and D. Munz, "Indentation cracking of brittle thin films on brittle substrates," in *Fracture Mechanics of Ceramics*, Vol. 12, R.C. Bradt et al. eds., Plenum Press, N.Y., 1996.

Appendix 1
Elastic Indentation Stress Fields

The methods of analysis of nanoindentation test data rely heavily on the elastic unloading response of the system. It is of interest therefore to have some appreciation of the equations for elastic contact and the associated indentation stress fields. The following assumptions are an essential component of the analytical equations:

- The radii of curvature of the contacting bodies are large compared with the radius of the circle of contact.
- The dimensions of each body are large compared with the radius of the circle of contact. This allows indentation stresses and strains to be considered independently of those arising from the geometry, method of attachment, and boundaries of each solid.
- The contacting bodies are in frictionless contact. That is, only a normal pressure is transmitted between the indenter and the specimen.

A1.1 Contact Pressure Distributions

The indentation stress fields arise from the pressure distributions shown in Table A.1, which are applied over a contact radius on the surface of a linearly elastic semi-infinite half-space.

Table A.1.1 Equations for surface pressure distributions beneath the indenter for different types of indenters.

Indenter type	Equation for normal pressure distribution $r < a$
Sphere	$\dfrac{\sigma_z}{p_m} = -\dfrac{3}{2}\left(1 - \dfrac{r^2}{a^2}\right)^{1/2}$
Cylindrical flat punch	$\dfrac{\sigma_z}{p_m} = -\dfrac{1}{2}\left(1 - \dfrac{r^2}{a^2}\right)^{-1/2}$
Cone	$\dfrac{\sigma_z}{p_m} = -\cosh^{-1}\dfrac{a}{r}$

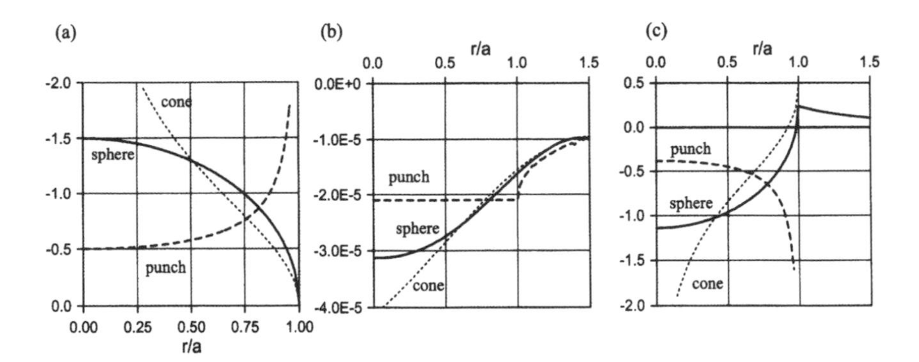

Fig. A1.1 (a) Normalized contact pressure distribution σ_z/p_m for spherical indenter, cylindrical punch, and conical indenters. (b) Deflection of the surface spherical, cylindrical, and conical indenters. Deflections in mm calculated for $p_m = 1$ MPa and radius of circle of contact = 1 mm and for E = 70 GPa. (c) Magnitude of normalized surface radial stress σ_r/p_m for spherical, cylindrical, and conical indenters. Calculated for $\nu = 0.26$ (after reference 1).

The pressure distributions, normal displacements and the magnitude of the radial stresses associated with an elastic contact with these indenters are shown in Fig. A1.1.

A1.2 Indentation Stress Fields

A1.2.1 Spherical Indenter

An equation for the normal pressure distribution directly beneath a spherical indenter was given by Hertz and is shown in Table A.1. As shown in Fig. A1.1a, the normal pressure $\sigma_z = 1.5p_m$ is a maximum at the center of contact and is zero at the edge of the contact circle. Outside the contact circle, the normal stress σ_z is zero, it being a free surface. The displacement of points on the surface of the specimen within the contact circle, measured with respect to the original specimen free surface, is given by:

$$u_z = \frac{1-\nu^2}{E}\frac{3}{2}p_m\frac{\pi}{4a}\left(2a^2 - r^2\right) \quad r \le a \tag{A1.2.1a}$$

Note that for all values of r < a, the displacement of points on the surface is inward toward the center of contact. Within the interior of the specimen, the stresses have a distribution depicted in Fig. A1.2. The contours shown in Figs.

A1.2 (a) to (e) give no information about the direction or line of action of these stresses. Such information is only available by examining stress trajectories. Stress trajectories are curves whose tangents show the direction of one of the principal stresses at the point of tangency and are particularly useful in visualizing the directions in which the principal stresses act. The stress trajectories of σ_2, being a hoop stress, are circles around the z axis. Stress trajectories for σ_1 and σ_3 are shown in Figs. A1.2 (f).

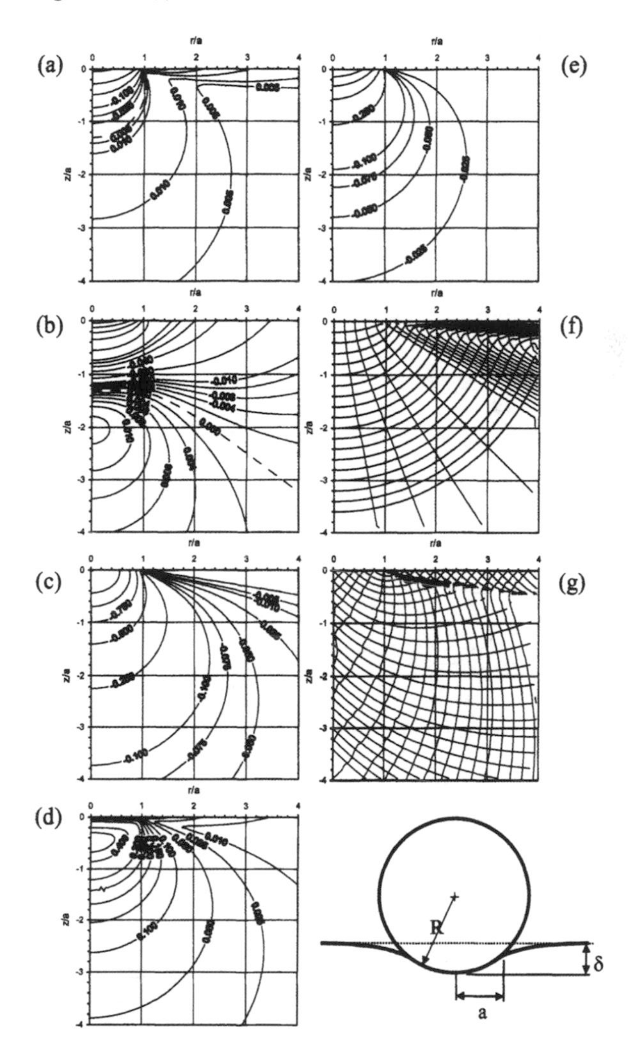

Fig. A1.2 Stress trajectories and contours of equal stress for spherical indenter calculated for Poisson's ratio $\nu = 0.26$. Distances r and z normalized to the contact radius a and stresses expressed in terms of the mean contact pressure p_m. (a) σ_1, (b) σ_2, (c) σ_3, (d) τ_{max}, (e) σ_H, (f) σ_1 and σ_3 trajectories, (g) τ_{max} trajectories (after reference 1).

A1.2.2 Conical Indenter

The mean contact pressure for a conical indenter depends only on the cone angle. The displacement beneath the original specimen surface is given by:

$$u_z = \left(\frac{\pi}{2} - \frac{r}{a}\right) a \cot\alpha \quad r \leq a \tag{A1.2.2a}$$

Contours of equal stress and stress trajectories for a conical indenter are shown in Fig. A1.3.

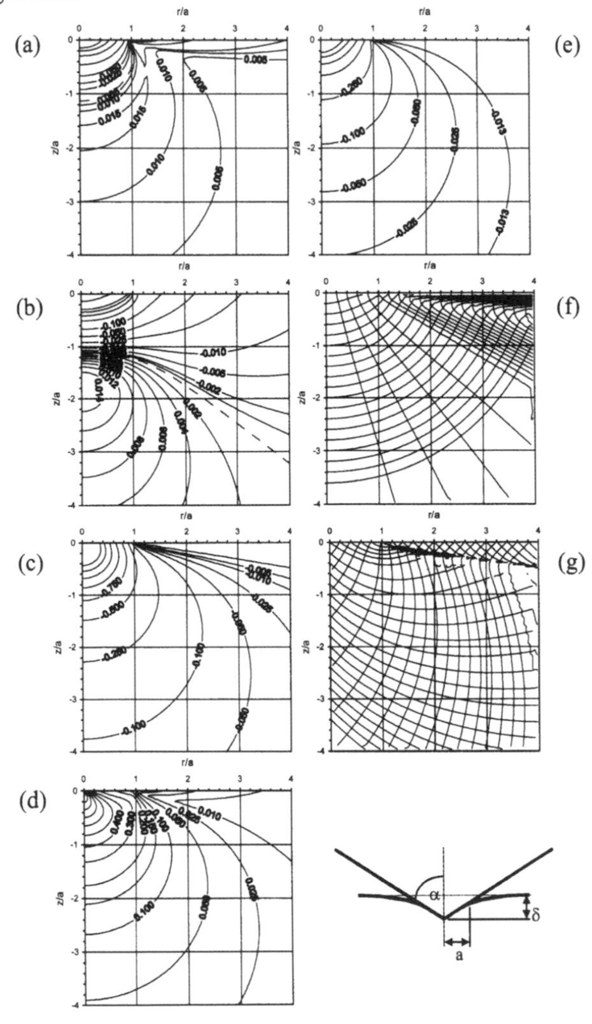

Fig. A1.3 Stress trajectories and contours of equal stress for conical indenter calculated for Poisson's ratio ν = 0.26. Distances r and z normalized to the contact radius a and stresses expressed in terms of the mean contact pressure p_m. (a) σ_1, (b) σ_2, (c) σ_3, (d) τ_{max}, (e) σ_H, (f) σ_1 and σ_3 trajectories, (g) τ_{max} trajectories (after reference 1).

References

1. A.C. Fischer-Cripps, *Introduction to Contact Mechanics*, Springer-Verlag, New York, 2000.

Appendix 2
Surface Forces, Adhesion, and Friction

A2.1 Adhesion Forces in Nanoindentation

With the increasing popularity of nanoscale technology, and the increasing sensitivity of nanoindentation instruments, it is appropriate to consider the effect of surface forces and adhesion in nanoindentation testing. The treatments of analysis presented in this book assume that there are no adhesive forces involved in the contact. That is, when the load is reduced to zero, the radius or dimension of the contact area also goes to zero. For contacts between very smooth surfaces, this is not the case owing to surface adhesion forces. In this appendix, the basic principles and significance of these types of forces in nanoindentation testing is introduced.

A2.2 Forces in Nature

It is generally agreed that there are four fundamental forces in nature. The so-called "strong" and "weak" nuclear forces are those found to exist over a very short range between neutrons, electrons, and protons and do not concern us here. The other two forces are the gravitational force and the electromagnetic force. It is the electromagnetic force that determines the nature of the physical interactions between molecules and is of most interest to us in this book. The electromagnetic force and the gravitational force often act together to provide physical phenomena that we are familiar with on a macroscopic scale.

Interactions (attractive and repulsive forces) between molecules take place over a relatively short distance scale of usually no more than 100 nm or so. However, the sum total effect of these short range interactions often lead to long-range macroscopic effects, e.g., the capillary rise of liquid in a narrow tube. It is important to note that the actual macroscopic physical properties of a liquid or a solid are determined by these relatively short-range interactions and do not generally depend on the size and shape of the solid or the container in the case of a liquid.

A2.3 Interaction Potentials

Intermolecular forces can generally not be fully described by a simple force law, such as the Universal Law of Gravitation The forces between molecules are influenced by the chemical nature of the elements involved, and also the proximity of neighboring molecules, whether they be of the same type or not. The combination of intermolecular forces from a variety of causes may result in a net attraction or repulsion between two molecules. It is sometimes convenient to talk of the molecular interactions in terms of their potential energy. Two widely separated molecules are said to have a high potential energy if there is an attractive force between them. If two such molecules are allowed to approach each other, the attractive force between them will cause them to accelerate and their initial potential energy is converted into kinetic energy. When these two molecules come together at an equilibrium position, this kinetic energy may be dissipated as heat and the potential energy of the interaction is at a minimum — a chemical bond has formed.

It is convenient to assign a potential energy of zero to widely spaced molecules so that when they approach and settle at their equilibrium position, the potential energy is a negative quantity. Since we generally assign a positive number to work done on a system (energy entering a system) and a negative number to work done by a system (energy leaving the system), work has to be done on the molecules to separate them.

Now, because of this assignment of signs to energy, we must be careful to assign the correct sign to intermolecular forces. It is common to define a positive force representing an attraction between two molecules. However, it turns out to be more convenient to assign negative forces as being attractive and positive forces as being repulsive. When multiplied by an appropriate distance, the energy change, positive or negative, will be then correctly assigned.

A very common interaction potential is the Lennard-Jones potential[1] between two atoms. It takes the form:

$$w(r) = -\frac{A}{r^6} + \frac{B}{r^{12}} \tag{A2.3.1}$$

where r is the distance between the atoms and the negative term represents the energy associated with attractive forces and the positive term represents that associated with the repulsive forces. The energies and forces associated with the Lennard-Jones potential are shown in Fig. A2.1.

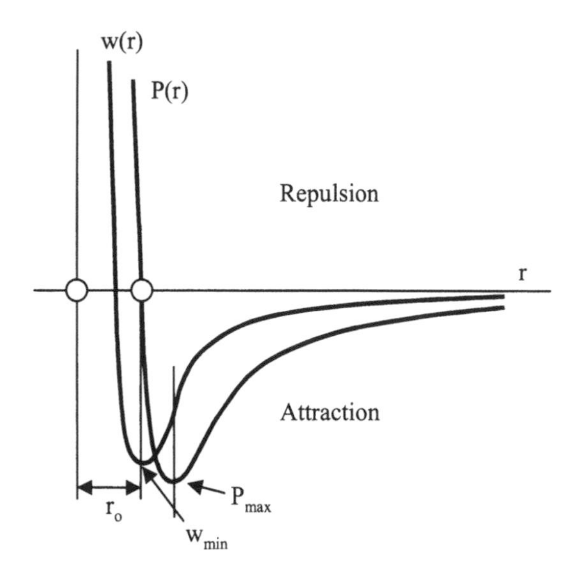

Fig. A2.1 The Lennard-Jones potential and interatomic forces between two atoms. The interaction force P(r) is zero where the energy w(r) is at a minimum, and the interaction force is a maximum when the rate of change of energy with respect to the distance between the two atoms is a maximum.

Note that in Fig. A2.1, the interaction force P(r) is zero where the energy w(r) is at a minimum, and the interaction force is a maximum (attractive in this case) when the rate of change, of energy with respect to the distance between the two atoms, is a maximum. Since we are mainly concerned with the significance of surface forces on adhesion and friction between solids in contact, we will find that we are dealing with the relatively long-range attractive forces.

A2.4 Van der Waals Forces

The attractive forces between atoms and molecules are known as the van der Waals forces after the scientist who studied them in 1873. van der Waals forces encompass electrostatic interactions that might arise from attractions between polar molecules, attractions between polarized molecules, and what are known as dispersion, or "London," forces. The dispersion forces, unlike the others mentioned, act between all molecules and are thus the most important component of the total van der Waals interactions between molecules. They may be attractive or repulsive and arise due to quantum mechanical effects. Very simply, they arise due to the instantaneous fluctuation in dipole moment, or distribution of charge, within an atom. Such instantaneous dipoles may polarize a nearby atom, thus causing a mutual attraction or repulsion. Dispersion forces can be quite substantial and can account for a large part of the intermolecular interactions of interest to us in this chapter.

Thus far we have considered interaction potentials, which invariably depend on distance. The associated forces P(r) between atoms and molecules can readily be found by differentiation, thus, for the case of the Lennard-Jones potential, we have:

$$\begin{aligned} P(r) &= -\frac{dw}{dr} \\ &= 6\frac{A}{r^7} - 12\frac{B}{r^{13}} \end{aligned} \qquad \text{(A2.4.1)}$$

van der Waals forces bind atoms or molecules together to form solids and liquids in many materials and are called "physical" bonds to distinguish them from chemical bonds — covalent and ionic bonds that act over very short distances.

A2.5 Surface Interactions

Intermolecular forces can be classified as being short range — operating over distance of less than a nanometre, and long range — operating up to distances of about 100 nm after which they are of negligible practical importance. The physical properties of gases are chiefly governed by interactive forces from neighboring atoms or molecules operating over relatively large distances. By contrast, the physical properties of solids are determined by intermolecular binding forces operating over relatively small distances. It should be remembered that all these interactions are electrostatic in origin, but their nature, on a molecular scale, is dictated by the cumulative effect of the proximity of neighboring atoms and the distance scale over which they operate. For example, a very important consequence of the effect of nearby ions of opposite sign in an ionic substance causes a screening effect that reduces the inverse square dependence of the Coulomb force so that the fall off with distance is more rapid.

For surfaces of bodies in contact, the situation is different again. It is found that the interaction potential between two bodies depends upon their absolute size and falls off much more slowly with increasing distance of separation compared to the interaction between two single molecules. The nature of the interaction potential may be quite complex, containing energy that prevents two molecules from attaining a minimum potential energy.

It is of interest to determine the surface interactions between two parallel plane surfaces since this offers a baseline for comparison with other surface interactions. It can be shown by calculating the interaction potential of one molecule on one surface with all the molecules on the other surface and then summing for all the molecules on the first surface, the total interaction potential for the case of van der Waals forces becomes[2]:

$$w(r)_{planes} = -\frac{\pi C \rho^2}{12r^2} \tag{A2.5.1}$$

where ρ is the number density of molecules within the two surfaces and C is a constant. For a large rigid sphere of radius R in close proximity to a flat rigid surface, the interaction potential can be derived in a similar way and is[2]:

$$w(r)_{sphere-plane} = -\frac{\pi^2 C \rho^2 R}{6r} \tag{A2.5.2}$$

Comparison with Eq. A2.3.1 shows that the interaction potentials for aggregated bodies decay at a substantially lower rate than that normally associated with van der Waals forces between two isolated atoms or molecules.

The associated expressions for the forces for the above two cases can be readily determined by differentiation. For the case of two plane surfaces, we obtain from Eq. A2.5.1:

$$P(r)_{planes} = \frac{dw}{dr} = -2\frac{\pi C \rho^2}{12r^3} \tag{A2.5.1}$$

And for the case of a sphere and a flat surface, differentiation of Eq. A2.5.2 gives:

$$\begin{aligned} P(r)_{sphere-plane} &= \frac{dw}{dr} = -\frac{\pi^2 C \rho^2 R}{6r^2} \\ &= -[2\pi R]\frac{\pi C \rho^2}{12r^2} \\ &= [2\pi R]w(r)_{planes} \end{aligned} \tag{A2.5.2}$$

Equation A2.5.2 expresses the surface forces between a sphere and a plane surface in terms of the interaction potential of two plane surfaces for the case of small separations (i.e., $r << R$).

For the case of two rigid spheres of radii R_1 and R_2, it can be shown that the surface force between them is[3]:

$$P(r)_{sphere-sphere} = [2\pi R]w(r)_{planes} \quad r << R_1, R_2 \tag{A2.5.3}$$

where R is the relative radii of curvature of the two spheres (see Eq. 1.2c). Equation A2.5.3 is known as the Derjaguin approximation.[4] The advantage of expressing surface forces in this way is that it is easy to determine interaction potentials between two planar surfaces and this potential can be then be used to determine the forces associated with more complicated geometrical surfaces.

An important practical result arising from the Derjaguin approximation is the force law associated with two crossed cylinders of radii R_1 and R_2. The resulting expression is:

$$P(r)_{\text{cylinder-cylinder}} = \left[2\pi\sqrt{R_1R_2}\right]w(r)_{\text{planes}} \quad r << R_1, R_2 \quad \text{(A2.5.4)}$$

When $R_1 = R_2$, we very conveniently obtain Eq. A2.5.2 — an important consequence for users of surface force apparatus (SFA) consisting of two crossed cylinders of equal radii. That is, contact between two crossed cylinders of equal radii is equivalent to that between a sphere of the same radius and a plane.

A2.6 Adhesion

We are now in a position to discuss the change in potential energy of a system when two surfaces of the same material bought together into contact in a vacuum. When the two surfaces come together, there is a negative potential energy ΔW associated with the contact. This energy is called the work of cohesion of the interface. (If the two surfaces are of the different materials, the energy is called the work of "adhesion.") If the material is then cleaved and the two halves separated, then work, equal to the work of cohesion, is done on the system and appears as "surface energy" of the two new surfaces, thus:

$$-\Delta W = 2\gamma \quad \text{(A2.6.1)}$$

If two surfaces of a material 1 are initially in a medium 2 and then bought into contact, then it can be shown that the energy associated with the interface is given by:

$$\gamma = \gamma_1 + \lambda_2 - 2\sqrt{\gamma_1\gamma_2} \quad \text{(A2.6.2)}$$

In Eq. A2.6.2, γ is half of the total energy required to separate two surfaces of material 1 immersed in a medium 2.§§§

For the case of two rigid spheres in contact, the adhesion force P_A is given by the Derjaguin approximation, where we can now express the force in terms of the unit surface energy of the contact of two planar surfaces (See Eq. A2.5.3). Expressing the adhesion force as that required to separate the two spheres of relative radii R, we have[3]:

$$P_A = -4\pi R\Delta\gamma \quad \text{(A2.6.3)}$$

Equation A2.6.3 applies to the case of rigid spheres in contact — i.e., the contact radius is zero. In practice, the spheres deform when placed in contact due to their finite value of elastic modulus. Under zero applied external load, two non-rigid spheres placed in contact will deform locally because of adhesion

§§§ It should be noted that many texts define γ as the "interfacial surface energy" equal to the total energy required to separate two surfaces. Here, we define γ as the surface energy associated with each of the two new surfaces that is half of the interfacial surface energy. We have left the 2γ terms in these equations to remind ourselves of the distinction.

forces that produce a finite radius of circle of contact. Derjaguin, Muller, and Topov[5] accounted for the case of deformable bodies by adding the force given by Eq. A2.6.3 to the Hertz contact equations and the resulting contact is referred to as the "DMT" theory. The contact deformation remains Hertzian and the adhesion force P_A acts in addition to the externally applied force:

$$P = \frac{4E^*}{3}\frac{a^3}{R} - 4\pi R\Delta\gamma \tag{A2.6.4}$$

Following Johnson, Kendall, and Roberts,[6] one could take that view that the finite size of the contact area is equivalent to a subsequent loss of surface energy $U_S = -2\gamma\pi a^2$ as the spheres come into contact. The contact radius grows in size until a balance is reached between the loss of surface energy and the increase in stored elastic energy at the deformed material in the vicinity of the contact:

$$\frac{dU_E}{da} = \frac{dU_S}{da} = -4\gamma\pi a \tag{A2.6.5}$$

Now, the Hertz pressure distribution p(r) can by rewritten as:

$$p(r) = -\frac{3}{2}\frac{P}{\pi a^2}\left(1 - \frac{r^2}{a^2}\right)^{1/2} \quad r \le a \tag{A2.6.6}$$

where the minus sign indicates compression.

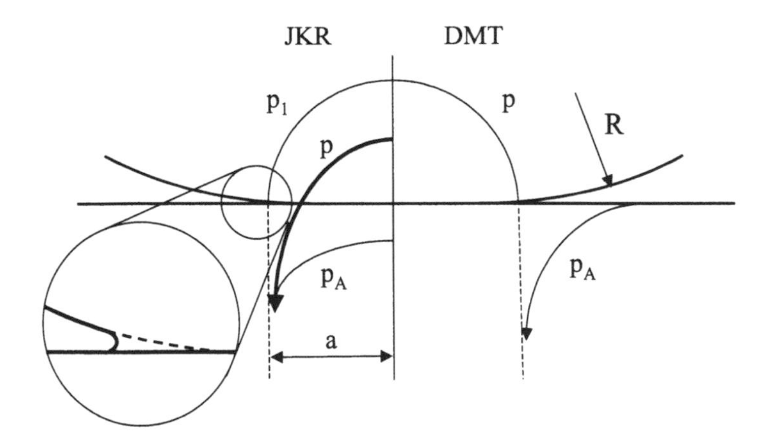

Fig. A2.2 Contact conditions for a deformable sphere under external load P on a rigid flat surface. In the JKR case, p is the Hertz contact pressure distribution, p_A is the tensile stress due to adhesion, and p_1 is the actual contact pressure distribution, which is the sum of p and p_A. The contact radius a is larger than that predicted by the Hertz equations alone. The corresponding features of the DMT theory are also shown where the stresses due to adhesion forces act outside the radius of the circle of contact. The contact angle depends upon the relative surface energies of the contacting bodies.

The pressure distribution (tension) due to adhesion forces is shown by Johnson[7] to be:

$$p_A(r) = \frac{P}{2\pi a^2}\left(1 - \frac{r^2}{a^2}\right)^{-1/2} \qquad \text{(A2.6.7)}$$

These two pressure distributions and their sum are shown in Fig. A2.2 as the JKR theory along with the corresponding features of the DMT theory. The net pressure distribution results in a stored elastic strain energy given by:

$$p_A(r) = \frac{P}{2\pi a^2}\left(1 - \frac{r^2}{a^2}\right)^{-1/2} \qquad \text{(A2.6.8)}$$

and taking the derivative with respect to a, and equating with the energy balance in Eq. A2.6.3, we find that the force due to adhesion for an external load that produces a total contact of radius a is[8]:

$$P_A = \sqrt{8\pi a^3 2\gamma E^*} \qquad \text{(A2.6.9)}$$

Adhesion forces increase the contact radius for a particular applied load P over that predicted by the Hertz equations. The apparent load P_1 acting between two surfaces in which adhesion is acting is thus $P_1 = P + P_A$, or:

$$\begin{aligned} P &= P_1 - P_A \\ &= \frac{4}{3}E^* \frac{a^3}{R} - \sqrt{8\pi a^3 2\gamma E^*} \end{aligned} \qquad \text{(A2.6.10)}$$

where a is the actual radius of the circle of contact, greater than that predicted by the Hertz equations. For a given applied load P, the actual radius of the circle of contact is given:

$$a^3 = \frac{3}{4}\frac{R}{E^*}\left[P + 3\pi R 2\gamma + \sqrt{6\pi R 2\gamma P + (3\pi R 2\gamma)^2}\right] \qquad \text{(A2.6.11)}$$

When P = 0, the contact radius is a_o and from Eq. A2.6.11 is found to be:

$$a_o{}^3 = \frac{3}{4}\frac{6\pi R^2 2\gamma}{E^*} \qquad \text{(A2.6.12)}$$

and the adhesion force P_o acting to keep the spheres in contact at this condition is:

$$P_o = 6\pi R 2\gamma \qquad \text{(A2.6.13)}$$

For the case of $\gamma = 0$, Eq. A2.6.11 reduces to the Hertz equation. It should be noted that P_o is not the force required to separate the two spheres. If a negative load P is applied, then as long as the term inside the square root in Eq. A2.6.11

remains positive, the spheres remain in contact with an ever decreasing contact radius until a critical negative load P_C is reached, at which time the spheres abruptly separate.

$$\begin{aligned} P_C &= -\frac{3}{2}\pi R 2\gamma \\ &= -3\pi R\gamma \end{aligned} \qquad \text{(A2.6.14)}$$

It should be noted that the pull-off force P_C is independent of the elastic modulus but only depends on the relative radii of curvature and the surface energy. The significance is that Eq. A2.6.14 should apply equally well to rigid spheres, but this would be contradictory to Eq. A2.6.3. The apparent conflict was resolved by Tabor who proposed that the two theories represented the opposite extremes of a dimensionless parameter μ given by:

$$\mu = \left(\frac{4R\gamma^2}{E^{*2} z_o^3} \right)^{1/3} \qquad \text{(A2.6.15)}$$

In Eq. A2.6.15, z_o is the equilibrium spacing in the Lennard-Jones potential. The significance of Eq. A2.6.15 is that it represents the ratio of the elastic deformation due to adhesion to their range of action. The JKR theory (Eq. A2.6.10) is applicable to large radius compliant solids (μ >5) and the DMT theory applies to small rigid solids (μ<0.1). Physically, the JKR theory accounts for adhesion forces only within the expanded area of contact, whereas the DMT theory accounts for adhesion forces only just outside the contact circle. The intermediate regime has been extensively studied[9–12] but for most practical applications, the JKR theory applies.

Taking the ratio of P_1 and P_A in Eq. A2.6.9, we find:

$$\frac{P_1}{P_A} = \frac{1}{3}\frac{a}{R}\frac{\sqrt{E^* a}}{\sqrt{\pi\gamma}} \qquad \text{(A2.6.16)}$$

Now, the quantity a/R is the indentation strain and sets the scale of the contact. For a particular contact, then, the value of E^*a determines whether or not the adhesion force is significant. As E^*a becomes smaller, the adhesion force P_A becomes larger. Thus, the adhesive force is significant for very compliant surfaces even when the contacts are large. For the case of large E^*, the adhesion force becomes significant at very small contacts.

The load-point displacement when accounting for adhesive contact is also greater than that predicted by the Hertz equations and is found to be:

$$\delta = \frac{a^2}{R}\left(1 - \frac{2}{3}\left(\frac{a_o}{a}\right)^{3/2}\right) \qquad \text{(A2.6.17)}$$

The influence of surface roughness upon the adhesive forces acting between two solids has been studied by Fuller and Tabor[13] and also by Maugis[14]. In practice, surfaces consist of a range of asperity heights that may deform elastically or plastically when pressed together. As these surfaces separate, the junctions between the lower asperities are progressively broken by the relaxation of the compressive stresses in the higher ones until only a few junctions remain. The pull-off force drops rapidly with increasing surface roughness.[15]

A2.7 Friction

Amontons' laws of (dry) friction, in which it is stated that the resistance to motion between two bodies in contact is proportional to the load and the nature of the contacting surfaces is independent on the area of contact, were formulated a little over 300 years ago. These famous laws are of considerable practical importance in engineering applications and are most usually expressed in terms of a coefficient of friction, μ, such that:

$$F = \mu N \tag{A2.7.1}$$

where N is the applied normal load and F is the sideways force required to initiate tangential sliding between the two bodies. Once the two bodies in question are sliding past one another, it is found that the coefficient of friction is usually reduced. It should be recognized that the real area of contact between two bodies is usually less than the apparent area of contact due to the inherent roughness of real surfaces. Contact thus only takes place at asperities on each surface and friction forces arise from elastic deformation, shearing, welding, ploughing, or plastic deformation of the asperities as they move, or attempt to move, past one another. If we divide the terms in Eq. A2.7.1 by the real area of contact, we find that the frictional shear stress is in direct proportion to the normal stress:

$$\begin{aligned}\tau &= \frac{F}{A} = \frac{\mu N}{A} \\ &= \mu\sigma\end{aligned} \tag{A2.7.2}$$

If the frictional shear strength of the contact is a property of the contacting surfaces, then it should be independent of the applied normal stress. This means that, for Amontons' law to hold, the real area of contact A in Eq. A2.7.1 is directly proportional to the applied normal load N.

The Bowden and Tabor plastic junction theory[16] proposes that usually, the real area of contact between two surfaces is very small, and as a consequence, the contact pressure at those asperities in contact is very high. This results in plastic deformation of the asperities during contact so the real area of contact A can be calculated from the material's indentation hardness value H:

$$A = \frac{P}{H} \tag{A2.7.3}$$

where P is the applied normal force. Eq. A2.7.3 also shows the real area of contact is proportional to the applied normal force in accordance with Amontons' law.

Archard[17] subsequently proposed that, although the deformation of the asperities may be initially plastic, there is a steady-state condition in which the contact becomes completely elastic. Archard[17] showed that if, for elastic contact, the average size of the contacts remains constant, and an increase in load produces an increase in the total number of asperities coming into contact, then the proportionality between load and real area of contact is maintained. It is now generally agreed that Amontons' law, the proportionality between friction force and load, is independent of the nature of the contact at the asperities and is a consequence of the random roughness of real surfaces.[18]

A more fundamental knowledge of the phenomenon of friction and its relationship to adhesion is obtained through contacts in which the real and apparent areas coincide, thus removing the problems associated with surface roughness. In this type of contact the actual area of contact A is independent of the applied normal force (for a contact whose apparent area of contact does not change). Since the shear strength τ of a contact is a property of the contacting materials, then Amontons' law, Eq. A2.7.2 cannot apply. This means that the frictional shear strength of a contact, in which the real and apparent areas of contact coincide, is expected to the independent of the normal force.

Recent developments in this field have been made possible by the use of new instrumentation such as the surface force apparatus (SFA) and the atomic force microscope (AFM). In the SFA, contact is usually made between two crossed cylinders that have been lined with a thin sheet of cleaved mica. In the AFM, a very fine needle with a radius of approximately 10 nm or so is bought into close proximity to the surface to be measured and the resulting deflection of the cantilever, to which the needle or tip is attached, is used to measure the topography of the surface. In the atomic friction microscope, physical contact is made with the specimen surface and a tangential force is applied. The resulting twist of the cantilever is a measure of the friction force.

With these two instruments, a different value of frictional shear stress is measured for dry mica. In the AFM, the frictional shear stress is reported at about 1 GPa, whereas with the SFA, a frictional shear stress of about 20 MPa is found. Johnson[18] proposes that these differences in frictional stress (or shear strength) when measured with different types of instruments is due to the different physical mechanism associated with the scale of the contact. In the AFM, the scale or diameter of the contact is measured in nanometres. In the SFA, the scale of the contact is measured in micrometres. According to Johnson,[18] the physical mechanism of sliding friction is due to the nucleation and propagation of dislocation-like defects through the interface under the influence of the applied shear force. For small contacts (<20 nm), dislocation-like defects cannot be nucleated and atoms move relative to one another in unison with a shear strength equal to the theoretical strength of the material. For large contacts (>50 μm), the rate of nucleation of dislocation-like continuities at the leading edge of the contact becomes equal to the rate at which they disappear at the center and the frictional stress reaches a steady-state value. In the intermediate range, the shear stress required to nucleate dislocation-like defects at the leading edge of the contact depends upon the square root of the dimension of the contact in a manner similar to that of a Mode II crack.

The significance of this is that frictional effects within the microstructure of many materials may be influenced by the scale of their microstructural features which, in turn, dictate the mechanical strength of the solid in practical applications, especially in indentation loading.[19]

References

1. J.E. Lennard-Jones and B.M. Dent, "Cohesion at a Crystal Surface," Trans. Faraday Soc. 24, 1928, pp. 92–108.
2. J.N. Israelachvili, *Intermolecular and Surface Forces*, 2nd Ed. 1992, Academic Press, London.
3. R.S. Bradley, "The cohesive force between solid surfaces and the surface energy of solids," Phil. Mag. 13, 1932, pp. 853–862.
4. B.V. Derjaguin, "Theorie des Anhaftens kleiner Teilchen," Koll. Z. 69, 1934, pp. 155–164.
5. B.V. Derjaguin, V.M. Muller, and Y.P. Toporov, "Effect of contact deformations on the adhesion of particles," J. Coll. Interf. Sci. 53, 1975, pp. 314–326.
6. K.L. Johnson, K. Kendall, and A.D. Roberts, "Surface energy and the contact of elastic solids," Proc. Roy. Soc. A324, 1971, pp. 303–313.
7. K.L. Johnson, "Adhesion and Friction between a smooth elastic spherical asperity and a plane surface," Proc. Roy. Soc. A453, 1997, pp. 163–179.
8. K.L. Johnson, *Contact Mechanics*, Cambridge University Press, Cambridge, 1985.
9. J.A. Greenwood, "Contact of elastic spheres," Proc. Roy. Soc. A453, 1997, pp. 1277–1297.
10. V.M. Muller, V.S. Yushenko, and B.V. Derjaguin, "On the influence of molecular forces on the deformation of an elastic sphere and its sticking to a rigid plane," J. Coll. Interf. Sci. 77, 1980, pp. 91–101.
11. D. Maugis, "Adhesion of spheres: the JKR-DMT transition using a Dugdale model," J. Coll. Interf. Sci. 150, 1992, pp. 243–269.
12. N.A. Burnham and R.J. Colton, "Force Microscopy," in *Scanning Tunneling Microscopy and Spectroscopy: Theory, Techniques and Applications*, 1993, D.A. Bonnell, ed., VCH Publishers, New York.
13. K.N.G. Fuller and D. Tabor, "The effect of surface roughness on the adhesion of elastic solids," Proc. Roy. Soc. A324, 1975, pp. 327–342.
14. D. Maugis, "On the contact and adhesion of rough surfaces," J. Adhesion Sci. and Tech. 10, 1996, pp.161–175.

15. K.L. Johnson, "Mechanics of adhesion," Tribology International, 31 8, 1998, pp. 413–418.
16. F.P. Bowden and D. Tabor, *Friction and Lubrication of Solids,* Clarendon Press, Oxford, 1950.
17. J.F. Archard, "Elastic deformation and the laws of friction," Proc. Roy. Soc. A243, 1957, pp. 190–205.
18. K.L. Johnson, "The contribution of micro/nano-tribology to the interpretation of dry friction," Proc. Instn. Mech. Engrs. 214 C, 2000, pp. 11–22.
19. A.C. Fischer-Cripps, "The role of internal friction in indentation damage in a mica-containing glass-ceramic," J. Am. Ceram. Soc. 84 11, 2001, pp. 2603–2606.

Appendix 3
Common Indenter Geometries

A3.1 Berkovich Indenter

Berkovich Indenter

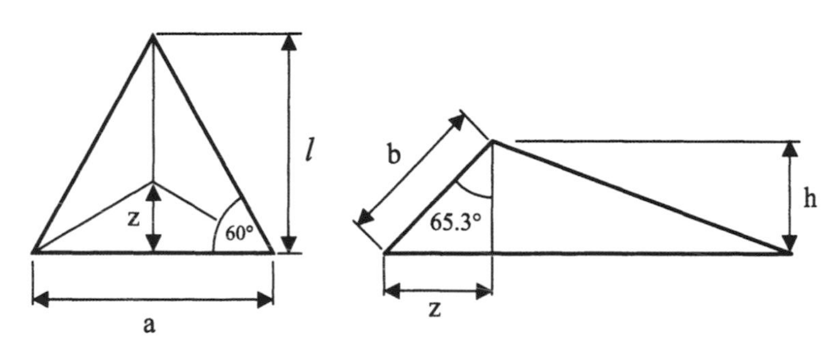

Projected area

$$\tan 60 = \frac{l}{a/2}$$

$$l = \frac{\sqrt{3}}{2}a$$

$$A_{proj} = \frac{al}{2}$$

$$= \frac{\sqrt{3}}{4}a^2 \approx 0.433a^2$$

$$\cos 65.27 = \frac{h}{b}$$

$$h = \frac{a\cos 65.27}{2\sqrt{3}\sin 65.27}$$

$$= \frac{a}{2\sqrt{3}\tan 65.27}$$

$$a = 2\sqrt{3}h\tan 65.27$$

$$A_{proj} = 3\sqrt{3}h^2\tan^2 65.27$$

$$= 24.49h^2$$

Surface area

$$A_{surf} = 3\frac{ab}{2}$$

$$\sin 65.27 = \frac{z}{b}$$

$$z = \frac{a}{2}\tan 30$$

$$= \frac{a}{2\sqrt{3}}$$

$$b = \frac{a}{2\sqrt{3}\sin 65.27}$$

$$A_{surf} = 3\frac{a^2}{4\sqrt{3}\sin 65.27}$$

$$\approx 0.477a^2$$

$$a = 2\sqrt{3}h\tan 65.27$$

$$A_{surf} \approx 26.98h^2$$

Equivalent cone angle:
70.3°

For cube corner, replace 65.27° with 35.264°

Equivalent cone angle: 42.28°

For original Berkovich indenter, 65.0333°

$A_{proj} = 23.97h^2$

$A_{surf} = 26.40h^2$

Fig. A3.1 Berkovich indenter.

A3.2 Vickers Indenter

Vickers Indenter

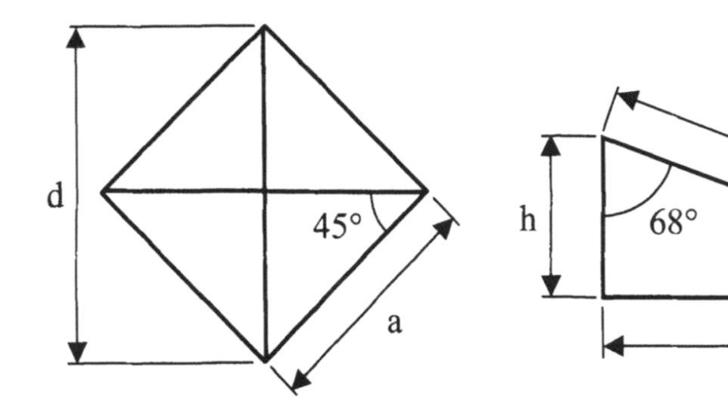

Projected area

$$\sin 45 = \frac{d}{2a}$$

$$a = \frac{d}{\sqrt{2}}$$

$$A_{proj} = a^2 = \frac{d^2}{2}$$

$$\tan 68 = \frac{a}{2h}$$

$$a = 2h \tan 68$$

$$A_{proj} = a^2 = 4h^2 \tan^2 68 \approx 24.504h^2$$

Surface area

$$A_{surf} = 4\frac{ab}{2}$$

$$\sin 68 = \frac{a}{2b}$$

$$b = \frac{a}{2 \sin 68}$$

$$A_{surf} = \frac{a^2}{\sin 68} = \frac{4h^2 \tan^2 68}{\sin 68} \approx 26.429h^2$$

Equivalent cone angle:

70.3°

Fig. A3.2 Vickers indenter.

A3.3 Knoop Indenter

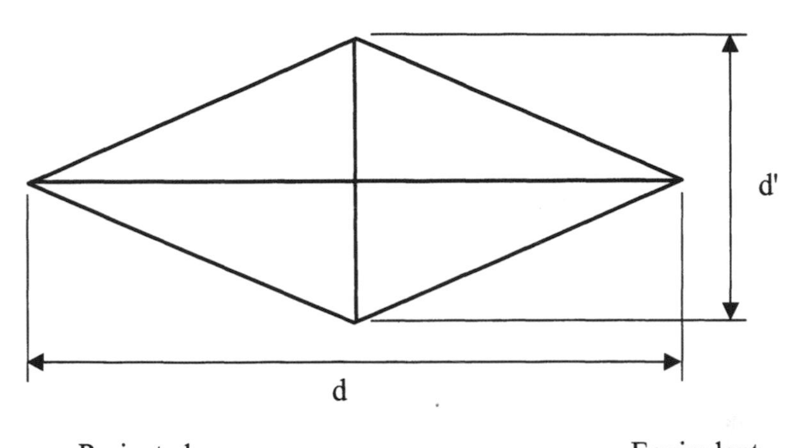

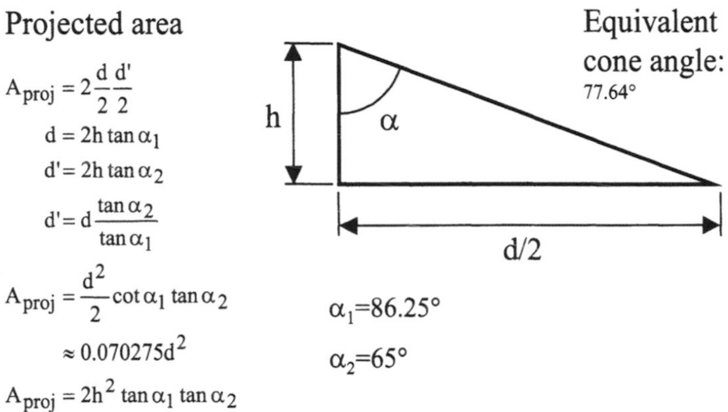

Fig. A3.3 Knoop indenter.

A3.4 Sphero-Conical Indenter

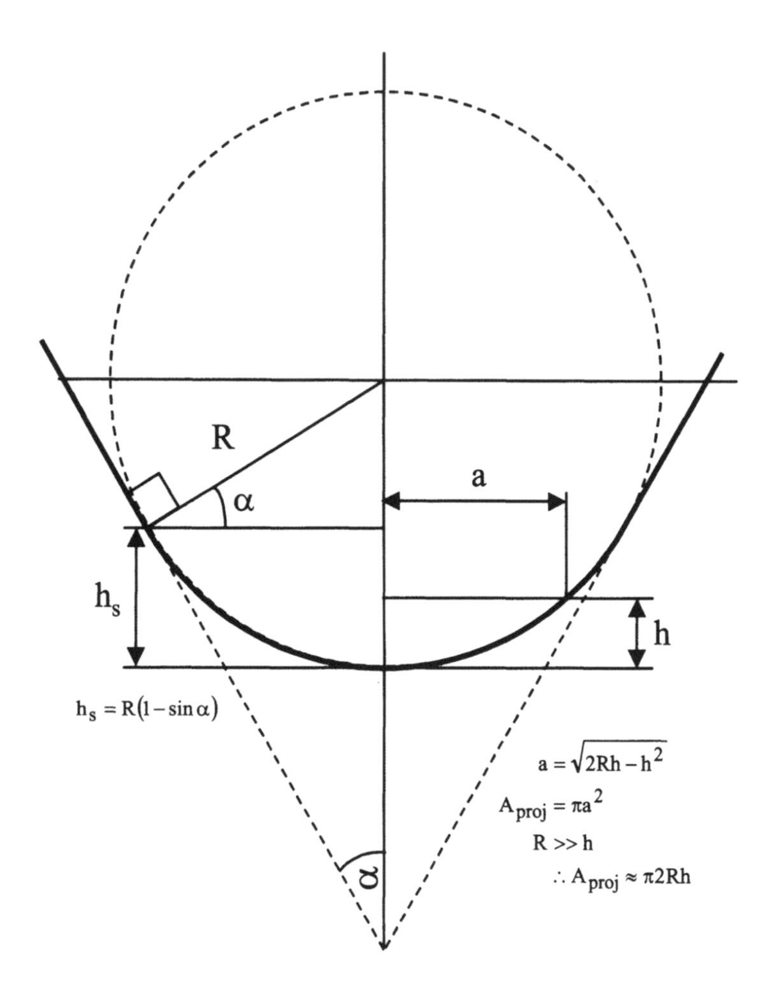

Fig. A3.4 Sphero-conical indenter.

Appendix 4
Non-linear Least Squares Fitting

In the analysis of nanoindentation load displacement curves, it is often required to fit a power law function to the experimental data. Such a fitting usually involves three or more independent variables. Similar types of analysis are required for the analysis of creep curves. In this appendix, a numerical method of fitting any number of independent parameters is given in some details.

Let Z_i be a function that provides fitted values of a dependent variable y_i at each value of an independent variable x_i. Z_i can be a function of many parameters a_o, a_1, … a_r.

$$Z_i = Z_i\left(x_i : a_o, a_1, a_2, \ldots a_j, \ldots a_r\right) \tag{A4.1}$$

It is presumed that initial values or estimates of these parameters are known and that the desired outcome is an optimization of the values of these parameters using the method of least squares. The true value of the parameter a_j is found by adding an error term δa_j to the initial value a_j^o.

$$a_j = a_j^o + \delta a_j \tag{A4.2}$$

Thus, the function Z_i becomes:

$$Z_i^o = Z_i\left(x_i : a_1^o, a_2^o, a_3^o \ldots a_r^o\right) \tag{A4.3}$$

If the errors δa_j are small, then the function Z_i can be expressed as a Taylor series expansion:

$$Z_i = Z_i^o + \sum_{j=1}^{r}\left(\frac{\delta Z_i^o}{\delta a_j}\right)\delta a_j \tag{A4.4}$$

This is a linear equation in δa_j and is thus amenable to multiple linear least squares analysis.

Now, by least squares theory, we wish to minimize the sum of the squares of the differences (or residuals) between the observed values y_i and the fitted values Z_i. The differences $y_i - Z_i$ at each data point i can be weighted by a factor w_i to reflect the error associated in the observed values y_i. Thus, the sum of the squares is expressed as:

$$X^2 = \sum_{i=1}^{N} w_i [y_i - Z_i]^2$$
$$= \sum_{i=1}^{N} w_i \left[y_i - \left(Z_i^o + \sum_{j=1}^{r} \left(\frac{\delta Z_i^o}{\delta a_j} \right) \delta a_j \right) \right]^2$$
$$= \sum_{i=1}^{N} w_i \left[\left(y_i - Z_i^o \right) - \sum_{j=1}^{r} \left(\frac{\delta Z_i^o}{\delta a_j} \right) \delta a_j \right]^2 \qquad \text{(A4.5)}$$
$$\Delta y_i = y_i - Z_i^o$$
$$X^2 = \sum_{i=1}^{N} w_i \left[\Delta y_i - \sum_{j=1}^{r} \left(\frac{\delta Z_i^o}{\delta a_j} \right) \delta a_j \right]^2$$

The weighting factor w_i for the present application can be simply the magnitude of y_i on the assumption that the error at each data point is inversely proportional to the magnitude of the data at that point.

The objective is to minimize this sum with respect to the values of the error terms δa_j, thus we set the derivative of X^2 with respect to δa_j to zero:

$$\frac{\delta X^2}{\delta(\delta a_j)} = 0 = \sum_{i=1}^{N} w_i \left[\left(\Delta y_i - \sum_{j=1}^{r} \left(\frac{\delta Z_i^o}{\delta a_j} \right) \delta a_j \right) \left(\frac{\delta Z_i^o}{\delta a_j} \right) \right] \qquad \text{(A4.6)}$$

This expression can be expanded by considering a few examples of j. Letting $j = 1$, we obtain:

$$\sum_{i=1}^{N} w_i \Delta y_i \left(\frac{\delta Z_i^o}{\delta a_1} \right) = \delta a_1 \sum_{i=1}^{N} w_i \left(\frac{\delta Z_i^o}{\delta a_1} \right)^2 + \delta a_2 \sum_{i=1}^{N} w_i \left(\frac{\delta Z_i^o}{\delta a_1} \frac{\delta Z_i^o}{\delta a_2} \right) + \\ \ldots. + \delta a_r \sum_{i=1}^{N} w_i \left(\frac{\delta Z_i^o}{\delta a_1} \frac{\delta Z_i^o}{\delta a_r} \right) \qquad \text{(A4.7)}$$

At j equal to some arbitrary value of k, we obtain:

$$\sum_{i=1}^{N} w_i \Delta y_i \left(\frac{\delta Z_i^o}{\delta a_k} \right) = \delta a_1 \sum_{i=1}^{N} w_i \left(\frac{\delta Z_i^o}{\delta a_k} \frac{\delta Z_i^o}{\delta a_1} \right) + \delta a_2 \sum_{i=1}^{N} w_i \left(\frac{\delta Z_i^o}{\delta a_k} \frac{\delta Z_i^o}{\delta a_2} \right) + \\ \ldots \delta a_k \sum_{i=1}^{N} w_i \left(\frac{\delta Z_i^o}{\delta a_k} \right)^2 + \ldots + \delta a_r \sum_{i=1}^{N} w_i \left(\frac{\delta Z_i^o}{\delta a_k} \frac{\delta Z_i^o}{\delta a_r} \right) \qquad \text{(A4.8)}$$

In matrix notation, the sums for each error term δa from 1 to r is expressed:

$$[Y_j] = [A_{jk}][X_j]$$

$$\begin{bmatrix} Y_1 \\ Y_2 \\ Y_k \\ . \\ Y_r \end{bmatrix} = \begin{bmatrix} A_{11} & . & . & . & A_{1r} \\ . & . & . & . & . \\ . & . & A_{jk} & . & . \\ . & . & . & . & . \\ A_{r1} & . & . & . & A_{rr} \end{bmatrix} \begin{bmatrix} X_1 \\ X_2 \\ X_k \\ . \\ X_r \end{bmatrix} \tag{A4.9}$$

where

$$Y_j = \sum_{i=1}^{N} w_i \left[\Delta y_i \frac{\delta Z_i^o}{\delta a_j} \right]$$

$$A_{jk} = A_{kj} = \sum_{i=1}^{N} w_i \left[\frac{\delta Z_i^o}{\delta a_j} \frac{\delta Z_i^o}{\delta a_k} \right] \tag{A4.10}$$

$$X_j = \delta a_j$$

and

$$\Delta y_i = y_i - Z_i^o \tag{A4.11}$$

The solution is the matrix X that contains the error terms to be minimized. Thus:

$$[X_j] = [A_{jk}]^{-1} [Y_j] \tag{A4.12}$$

When values of δa_j are calculated, they are added to the initial values a_j^o to give the fitted values a_j:

$$a_j^1 = a_j^o + L\delta a_j \tag{A4.13}$$

The process may then be repeated until the error terms δa_j become sufficiently small indicating that the parameters a_j have converged to their optimum value. L in Eq. A4.13 is a relaxation factor that can be applied to error terms to prevent instability during the initial phases of the refinement process.

A measure of the goodness of the fit is given by the correlation coefficient. This is the ratio of the sum of the squares of the differences between the fitted values Z_i and the average value of the y_is (i.e. the sum of the squares due to the regression line) to the sum of the squares of the differences between the actual data y_i and the average value.

$$r^2 = \frac{\sum (Z_i - \bar{y})^2}{\sum (y_i - \bar{y})^2} \tag{A4.14}$$

Values approaching unity indicate that the deviations from the mean value of all the y_is (or Z_is) at each data point arise mainly from the deviation from the regression line instead of deviations from the data point to the regression line. When all the data points lie on the regression line, $Z_i = y_i$.

In the case of application to nanoindentation testing, the following quantities are appropriate:

For fitting the unloading part of a load-displacement curve, the following expressions are appropriate:

$$\begin{aligned} Z_i &= a_1 (x_i - a_2)^{a_3} \\ a_1 &= A \\ a_2 &= h_r \\ a_3 &= n \end{aligned} \tag{A4.15}$$

According to the fitting procedure, the following functions are thus required for the analysis:

$$\begin{aligned} Z_i^o &= a_1^o \left(x_i - a_2^o\right)^{a_3^o} \\ \frac{\delta Z_i^o}{\delta a_1} &= \left(x_i - a_2^o\right)^{a_3^o} \\ \frac{\delta Z_i^o}{\delta a_2} &= -a_3^o a_1^o \left(x_i - a_2^o\right)^{a_3^o - 1} \\ \frac{\delta Z_i^o}{\delta a_3} &= a_1^o \left(x_i - a_2^o\right)^{a_3^o} \ln\left(a_1^o \left(x_i - a_2^o\right)\right) \end{aligned} \tag{A4.16}$$

Appendix 5
Properties of Materials

Table A.5.1 Experimental values of modulus and hardness for various types of thin films. (Measured at CSIRO Division of Telecommunications and Industrial Physics using a UMIS® instrument, data courtesy of A. Bendavid).

Film/Substrate	H (GPa)	E (GPa)
Si (substrate)	10–12	170–180
Ti	6.7–8.7	110–130
TiN	24–26	350–450
$TiSi_xN$	45–55	–
TiO_2 (amorphous)	11–13	130–150
TiO_2 (anatase)	11–12	160–180
TiO_2 (rutile)	18–19	200–220
Nb	6–7	–
NbN	38–42	–
NbC	40–45	–
Nb_2O_5	11–13	–
Al_2O_3	10–12	–
AlN	20–24	150–170
$AlSiN_x$	20–22	–
Ta_2O_5	9–10	130–150
DLC (CVD), + H_2	10–25	80–250
ta-C (fad), no H_2	50–60	500–600
ZrN	30–35	–
ZrO_2	11–12	–
Sapphire	28–32	380–400
Si_3N_4	18–24	180–250

Table A.5.2 Some properties of common specimen materials used in nanoindentation testing (Various sources).

Material	H (GPa)	E (GPa)	ν
Alumina	8.68		
Aluminium		70.5	0.34
Bronze		110	
Chromium		277	
Copper >99.95%	0.5	110	0.36
Copper Polycrystalline		130	0.343
Diamond		1070	0.07
Fused silica	8~10	72.5	0.17
Gold	0.3	80	0.44
Nickel		211	
Sapphire	30.0	400	0.21 – 0.27
Silicon	10~12	172	0.28
Silver		78	
Soda Glass	3~5	70	0.23
Tool steel	4~9	210	0.26

Appendix 6
Frequently Asked Questions

Nanoindentation testing is fairly straightforward, but the details can be overwhelming for the practitioner beginning in the field. Listed below are some frequently asked questions along with responses that might assist with an initial understanding of the issues involved.

1. What is thermal drift? How is it accounted for?

Thermal drift refers to a change in dimension of the indenter, specimen and the instrument resulting from a temperature change during the test. The depth sensor of a nanoindentation instrument is typically very sensitive with a resolution of less than a nanometre. For a constant applied load, any variation in depth sensor output is caused by either creep within the specimen material, or thermal drift. The most common reason for any such depth sensor output is thermal drift. Thermal drift is measured in nm/sec. A drift of only a few nanometres per second over a test cycle which might last a minute or so can introduce large errors into the load-displacement curve thus causing the measured modulus and hardness to be in error. Thermal expansion causing thermal drift is most pronounced at the contact between the indenter and the specimen. Here, the dimensions of the contact are very small (a few microns) and any expansion or contraction of the indenter tip and specimen surface is detected by the depth sensor. Because this contact takes place over very small dimensions, there is not so much of a thermal mass to act as a heat sink, so the contact responds very rapidly to changes in temperature.

There are two ways to handle thermal drift.

(a) Reduce the temperature variations at the specimen to an absolute minimum. This is the preferred method. This is accomplished by enclosing the instrument in a heavily insulated cabinet and locating the whole assembly in a temperature controlled laboratory. Variations in laboratory temperature of about 0.5 °C over an hour or so is usually sufficient to reduce thermal drift to a negligible level.

(b) Correct the data for thermal drift. This method is used if temperature variations cannot be eliminated due to the location of the instrument. The correction is performed by accumulating depth readings while holding the load constant. This hold period is usually performed over a 5 or 10 second period at either full load, or at the last data point on the unloading part of the test cycle. For thermal drift correction, hold data at unloading is usually preferred since any creep exhibited by the specimen (the effects of which are usually indistinguishable from thermal drift) are minimized. After this data has been collected, the thermal drift rate is established by fitting a straight line through the hold period using a least square fitting. The drift rate, in nm/sec, is then used to correct the depth sensor readings for the load-displacement data points adding or subtracting the product of the drift rate and the time at which the depth readings were taken, thus offsetting any effect of thermal drift. This procedure works reasonably well, but only if the drift rate is a constant value. If the temperature is either rising, or falling throughout the test, then this is usually an adequate correction. If the temperature changes rapidly so that there are both expansions and contractions at the contact during a test, then the correction will not be suitable.

2. The depth of the indentation is often recommended to be less than 10% the thickness of the film to minimize the influence of the substrate. Is it the same principle for measuring the hardness and modulus?

The commonly used "10% rule" is usually applied to the measurement of hardness but many people also apply it to the measurement of modulus. For hardness, the issue is that a proper measure of hardness can only be obtained with what is called a fully-developed plastic zone. All indenters, including nominally sharp indenters, are rounded at the very tip. For contact with a round or spherical indenter, the plastic zone begins beneath the specimen surface and gradually increases in size with increasing indenter load. During this period, the mean contact pressure (load divided by contact area) increases. At higher loads, the plastic zone becomes fully formed into an approximate hemispherical shape and the mean contact pressure approaches a constant value. Any increase in indenter load results in a proportional increase in contact area and the mean contact pressure remains the same. This value of mean contact pressure is called "hardness". Now, with a thin film, for hardness measurements, it is important that the development of the plastic zone be not influenced by the properties of the substrate. Thus, it is usually specified that for indentation depths of less than 10% of the film thickness, the plastic zone will not extend to a depth at which any deformation of the substrate has any effect. If the indenter load is too high, then the plastic zone may extend into the substrate (especially for a soft substrate) and so influence the value of hardness measured.

For modulus, the situation is quite different. Theoretically, any indentation, even at the lowest possible load, will result in some influence from the substrate since the elastic deflections of both the substrate and the film contribute to support the indenter load. Because of the localized nature of the indentation stress fields, more support comes from the film than the substrate, but there is always some influence. Hence, the 10% rule does not apply to modulus determinations although it is a reasonable place to begin a test for comparative purposes. The recommended procedure, should time permit, is to perform a series of indentations from a very low load to a reasonably high load and then plotting modulus vs indentation depth. Extrapolation of the data back to zero depth should result in a value of modulus for the film only.

3. What is compliance? How is it measured?

Compliance usually refers to the elastic compliance or stiffness of the indentation test instrument (although it often refers to the compliance of the indenter as well). When load is applied to the indenter, an equal and opposite reaction force is applied to the instrument load frame. The resulting deflection of the frame (usually in the order of nanometres) is registered by the depth sensor and thus, unless corrected for, introduces an error into the load-displacement curve obtained for a particular specimen. The stiffness of the load frame is taken to be a single value (i.e. a linear spring). The depth output resulting from the indenter reaction force is thus linearly dependent on the applied load. Thus, once the compliance of the instrument is known, the product of the compliance and the indenter load at each load increment can be subtracted from the depth readings so that the depth readings only refer to the localized deflections arising from the contact. It is desirable to have the compliance of the instrument be as low as possible so that the correction to be applied is not too large a proportion of the test data (i.e. high signal to noise ratio). The validity of the compliance correction relies solely on the determination of an accurate value of instrument compliance. A common fault is to arrive at a value of compliance for contact on say a fused silica specimen without checking to see that this value provides consistent values of modulus for stiffer materials (such as silicon or sapphire) for which the deflection of the load frame is a larger proportion of the overall depth signal (these latter materials having a large value of elastic modulus than fused silica).

It is important to note that the compliance referred to here does not apply to the localized deflection of the indenter due to the indentation into the specimen. This deflection is accounted for by the use of the "combined" or "reduced" modulus in the analysis procedure.

The value of compliance can be obtained by a number of methods. In one method, a series of indentations is made into a series of specimens for which the elastic modulus is well known. A plot of dh/dP vs $1/h_p$ yields a straight line whose intercept is the compliance of the instrument. In another method, the value of compliance can be measured directly by placing known weights onto the load frame of the instrument and measuring the resulting deflection.

4. For batch testing, what is the tolerance for the temperature variation during the testing?

A common feature of nanoindentation test instruments is the facility to perform multiple tests on either the same or different specimens. If no temperature correction or hold period is to be performed, then it is necessary to ensure that the temperature inside the environmental enclosure is kept stable to at least within 0.1 °C for any particular test. If a large number of tests are queued to be performed in unattended mode, then the range of temperatures that occur during the total testing time is not so important, as long as for each test, there is no more than about 0.1 °C change. The environmental enclosure usually supplied with an instrument will permit larger temperature changes to occur in the laboratory. Generally speaking, laboratory temperatures should not drift more than 5 °C per hour at the very most.

5. How is friction coefficient measured?

Friction coefficient is usually measured using a scratch tester option. Scratch testing is usually, but not necessarily, done by moving the specimen while the indenter is in contact with the specimen. The movement is usually accomplished using the specimen stage mechanism in conjunction with a lateral force sensor built into the stage assembly (or placed there as an accessory). The friction coefficient is found by dividing the lateral force by the normal force. The friction coefficient can be calculated with a ramped or steady value of normal force.

6. Can adhesion between the thin-film and substrate be measured?

Measuring adhesion is not a well-developed procedure since the method used depends upon the type of film/substrate system being studied. Nanoindentation instruments themselves generally do not provide an "adhesion" test as such. It is usually up to the operator to devise a suitable loading or choose a suitable indenter that will provide information about adhesion of a particular system. Further, the adhesion has to be quantified in some way as either a stress, or a maximum load, or fracture energy, etc. Some operators simply identify features or discontinuities on the load displacement curve as being indicators of adhesive failure.

7. How is the right load selected for scratch testing?

Like adhesion testing, scratch testing is not a well-developed procedure. There is no one measure of "scratch resistance". Some workers identify changes in friction coefficient as an indication of scratch resistance, some identify delamination (using images of the scratch) and a "critical load" as being a measure of scratch resistance. These features often depend on the material system being tested as well as the shape of the indenter, the duration and rate of application of load, etc. It is usual in these circumstances to attempt to reproduce the in-service conditions of loading in a controlled manner to evaluate scratch performance of material systems.

8. Is it possible to do nanoindentation measurements at high temperature?

This is difficult, but possible. For high temperature indentation testing, it is necessary to have a very stable thermal environment so as to minimize effects due to temperature changes. For example, it is usually necessary to ensure that both the indenter and the specimen are at the exact same temperature before contact so that when contact is made, one part does not cool or heat the other part thus causing an expansion or contraction of the contact. This can be done with a separate tip heater as well as the standard specimen heater. The heaters must be controlled using PID temperature controllers to ensure the minimum of thermal drift. Practically, the maximum temperature possible for this type of testing is limited by the thermal properties of the indenter. For a diamond indenter, temperatures up to about 600 °C are possible. Higher than this, a shielding atmosphere would be required to prevent oxidation of the diamond to carbon.

9. What is the general rule for choosing the right indenter for the different material of thin films and substrates?

In nanoindentation testing, there are several possibilities with regard to indenter shape. The most commonly used indenter shape is the three-sided pyramidal Berkovich indenter. This provides the sharpest point for hardness measurements on both thin films and bulk specimens. Spherical indenters also offer possibilities since the evolution of the plastic zone is a gradual occurrence and valuable information about the elastic and plastic properties of the specimen can be obtained from the one indentation test (e.g. phase changes, cracking, etc are usually more easily observed with a spherical indenter). For very thin films, where elastic modulus is the most important quantity to be measured, a spherical indenter is usually preferred since its shape can be more accurately determined. For very thin films where hardness is of prime concern, a Berkovich indenter is preferred because it results in a fully formed plastic zone at the lowest possible load. Spherical indenters are also useful for soft films, and for surfaces with significant surface roughness.

10. How can one determine if an indenter has been damaged?

In normal use, an indenter can last many years although sharp indenter naturally become blunt with use. The indenters are usually diamond ground to the required shape. The diamond material is hard-wearing, but brittle. An indenter can be chipped or broken if it is dropped or mishandled. In the normal course of usage, the shape or area function of the indenter is required to be measured at intervals appropriate to the number of indentations and types of material being indented. Usually, any problems with an indenter can be identified by indenting a standard material at a specified load at regular intervals. For example, 50 mN on fused silica with a Berkovich indenter usually results in an indentation depth of about 650 nm for a sharp indenter going down to about 580 nm for a blunt indenter. Should there be any sudden deviation from a measured value of this type, a full area function calibration of the indenter should be performed. SEM imaging of the indenter should be undertaken if it is suspected that the indenter has become chipped or broken.

11. What are typical resolutions for load and depth measurements for nanoindentation instruments?

The specifications for nanoindentation instruments are generally all very similar. Specifications of nanoindentation instruments vary according to the method used to determine them. Some manufacturers quote theoretical resolutions based upon the voltage range of the signals and the width of the analog to digital conversions done in the instrument. This provides a method of comparison between instruments, but means little when a real-world test is to be made. Practically, test results are limited by environmental factors such as fluctuations in ambient temperature and vibration rather than by instrument specifications. The non-identifying specifications given below are the theoretical figures and the "noise floor". The noise floor figures represent the best that the manufacturer expects to get under ideal laboratory conditions.

Table A6.1 Typical displacement specifications for a range of nanoindentation test instruments.

	#1	#2	#3	#4	#5
Max displacement	500 μm	5μm (up to 50 μm)	20 μm	100 μm	2 μm, 20 μm
Displacement resolution	0.02 nm	0.0002 nm	0.03 nm	Not available	0.003 nm
Displacement noise floor	Not available	0.2 nm	Not available	0.1 nm	0.05 nm

Table A6.2 Typical force specifications for a range of nanoindentation test instruments.

	#1	#2	#3	#4	#5
Max Force	500 mN	30 mN	300 mN	1 mN	50 mN, 500 mN
Force resolution	50 nN	1 nN	1 μN	Not available	50 nN
Force noise floor	Not available	100 nN	Not available	100 nN	300 nN

Most manufacturers would either employ a 16 bit, or 20 bit analogue to digital converter (ADC) in their systems and the theoretical resolution for each instrument can be determined by dividing the range (whether force or depth) by 2 raised to the power of the width of the ADC. For example, for a range of 50mN and a 16bit ADC, the theoretical resolution would be 50mN divided by 216 = 750nN. This figure can be further divided by a factor equal to the square root of the number of readings taken for averaging. The very small figures for resolution presented above are thus a combination of the smoothing effect of taking many readings and averaging the results and the width of the ADC and the range.

It should be noted that the noise floor of the specifications is the most important factor for the end user. Any increase in resolution beyond the noise floor will only mean that the noise is being measured more precisely. There is no value (other than claiming very high theoretical resolutions) in comparing theoretical figures for resolution if the usable resolution is limited by electronic noise within the instrument, environmental noise and vibration plus thermal drift. The minimum contact force is an important specification since this, amongst other things, determines the minimum thickness of thin film specimen that can be usefully tested.

12. How can the viscoelastic properties for organic materials be measured?

Conventional nanoindentation testing is for the purpose of measuring modulus and hardness of solids without regard to any viscous component in their mechanical properties. However, some materials, especially plastics and organic materials, are viscoelastic and exhibit creep or time-dependent behavior under load. It is possible to measure this time dependent behavior. There are two methods in general use. In one, a small oscillatory motion is applied to the indenter shaft during indentation. The force and depth signals, now being out of phase due to the viscous nature of the specimen, can be measured and an appropriate mechanical model (e.g. a spring and dashpot) used to analyse the resulting transfer function. This procedure requires careful calibration of the indentation apparatus since the dynamic effects of the instrument and its filter circuitry, will influence the readings taken in this way. In another method, a creep response is measured (similar to the thermal drift hold period) in which the load is held constant and the depth measured at regular intervals over a 10 to 20 second period. A spring and dashpot model can then be fitted to this creep data and viscoelastic components calculated. This second method has the advantage of not requiring dedicated AC signal and driver apparatus and also occurs at a low loading rate, thus eliminating dynamic effects due to instrument behavior.

13. What indentation rate, dwell time, and hold periods should be used for nanoindentation testing? Should these be different for different types of materials?

In some nanoindentation test instruments, the operator is allowed to alter various timings in the indentation cycle. For example, a dwell period before the start of each indentation can be programmed so that any mechanical vibrations from a previous specimen stage movement can be allowed to dissipate and not influence the test data to be collected. Also, a dwell period can be programmed for the instrument to wait before recording each load and depth readings at each load increment to allow for any dynamic effects (from the specimen or the instrument) to decay before the data is obtained. Hold periods at load and unload can also be programmed for thermal drift and creep measurements. Usually, a given set of dwell periods works well for most types of specimens. Some fine-tuning of these parameters may be useful for specimens with extreme mechanical properties or specimen mountings that are not ideal.

14. For brittle materials, should one always choose the smallest load possible?

Cracking within the specimen may affect the shape of the load-displacement curve obtained on brittle materials. However, it should be realized that the surface energy of such cracks are an extremely small proportion of the overall strain energy contained within the contact zone. Generally, the appearance of cracks, unless they are very large, do not affect the readings obtained using moderate loads in nanoindentation testing. For modulus measurements, a spherical indenter can also be used which minimizes the potential for cracking in these types of specimens.

15. What data is obtained from a scratch tester accessory?

A typical scratch tester accessory for a nanoindentation instrument allows the operator to program the scratch length, velocity, type of loading (constant, progressive or incremental). The resulting data usually consists of the normal force and displacement, the lateral force, and the specimen position. From this data, friction coefficient and the position of any discontinuities due to delamination can be identified.

16. How can a curved shape of specimen surface be tested?

The theoretical analysis embodied in nanoindentation testing usually presumes that the specimen surface is flat. However, surface curvature is readily accommodated since the depths of penetration and residual impression width are in the order of microns. If the curvature of the specimen is very small, and is known, then it is possible to account for this in the analysis by computing an effective indenter angle or radius. This is rarely done since most specimen curvature is large compared to the indentation size. Surface roughness however, is important since the local curvature of the specimen in this case is comparable to the indentation size. Special precautions are then required (such as the use of a spherical indenter or well-polished surface).

17. Are there any special precautions required for mounting the specimen?

Specimens for nanoindentation testing must be flat, well-polished, and securely mounted. It is usual to mount them on a flat steel specimen holder using hot wax or glue. Hot wax has the advantage of allowing the specimen to be easily removed from the holder after testing. For any type of specimen mounting, the overall desirable outcomes are for a flat surface (not sloping up or down – in case the specimen contacts the indenter prematurely during specimen movement) and a very secure mount (so that the compliance of the mounting does not affect the readings). For waxed or glued specimens, it is necessary to reduce the compliance of the wax layer by using the thinnest possible layer. The function of the wax should be to fill the voids on the underside of the specimen so that the specimen material itself is in contact with the specimen holder.

18. When the material exhibits significant creep how may an indentation test be performed?

Creep (as distinct from thermal drift) is usually evidenced by a load-displacement response that has a pronounced curvature during the unloading part of the indentation cycle. This results in the tangent to the initial unloading being negative. Data of this type can be handled using conventional unloading analysis as long as the creep rate diminishes with time to an acceptably low level and special precautions are taken. If creep is observed and the creep rate does not diminish with time, then the specimen is viscoelastic and special test procedures are required (see Question 12).

19. Is elastic recovery an issue with nanoindentation testing?

All materials (except for rigid–plastic materials) undergo elastic recovery when load is removed form the indenter. Indeed, it is this elastic recovery that leads to the unloading load-displacement curve and forms the basis of the analysis techniques that are used to calculate modulus and hardness of the specimen. For some materials, elastic recovery is complete (e.g. rubber) and in this sense, nanoindentation testing does not provide meaningful values of hardness or modulus. See also Question 22.

20. Can elastic energy and plastic energy during indentation be calculated?

Yes, the area enclosed by the load-displacement curve represents the energy lost through plastic deformation. The energy recovered elastically is the area underneath the elastic unloading curve. Some workers consider the plastic energy of deformation to be a measure of fracture toughness of the specimen, but this ignores any R-curve behavior associated with crack face bridging or other energy dissipative means.

21. What is initial penetration?

Nanoindentation is usually referred to as depth-sensing indentation because the technique usually involves the measurement of load and depth using instrument indentation instruments. The depth of penetration has to be measured from the specimen free-surface. In order to "zero" the depth sensor, it is necessary to bring the indenter into contact with the specimen surface at a very small initial contact load. When the initial contact load is reached, the depth sensor output is set to zero. This becomes the depth reference point for the load-displacement curve subsequently obtained. However, at the initial contact force, there is also a very small penetration. This is not accounted for in the depth readings so with respect to the specimen free surface, all the depth readings have to be increased a little to account for this initial penetration. This is done by fitting the load-displacement curve to a smooth polynomial and extrapolating to zero force. The resulting depth offset is then the "initial penetration" and is added to all the depth readings as a correction.

22. How does piling-up and sinking-in affect the test results?

Care has to be taken for materials that exhibit excessive piling-up or sinking-in around the indentation. For these types of material response, corrections are required to account for the extra support (in the case of piling-up) or lack of support (for sinking-in) to the indenter. These corrections are not well-characterized. They appear to be different for different types of materials. Piling-up, usually seen in some metals, results in a larger value of modulus and a lower value of hardness compared to those obtained by other methods. For this reason, the modulus obtained by nanoindentation experiments is sometimes referred to as the "indentation modulus" rather than the "elastic modulus".

23. Do phase-transformations within the specimen affect the indentation results?

Some materials exhibit phase transformations during nanoindentation testing as a result of the high values of hydrostatic stresses within the indentation stress field. This is most commonly observed in silicon specimens where a discontinuity is often observed in the unloading response. This may or may not affect the results of a nanoindentation test. For silicon, the upper portion of the unloading curve is used for fitting and subsequent calculations whereas the phase transformation in this material occurs at a lower load on the unloading curve.

24. Is there an optimum load to be used for different types of materials?

For bulk specimens, which exhibit no phase transformations and have no grading in properties with depth (i.e. surface modified layers), the load used for the indentation test does not affect the value of modulus obtained if the area function for the indenter is accurately determined. A good test of a nanoindentation instrument is to perform a series of indentations at increasing maximum loads and ensuring that the modulus is within an acceptable tolerance throughout.

For hardness measurements, the load must be large enough to overcome the initial elastic response even with a nominally sharp indenter. Hardness is usually taken to mean the mean contact pressure at a condition of a fully developed plastic zone.

25. Does residual stress in the specimen affect the results obtained? Can residual stress be measured?

Residual stresses do affect the shape of the load-displacement curve. This can be desirable in the sense that the properties of the specimen, complete with residual stress, are required to be measured. The level of residual stress can in some cases be measured but this is very difficult. Such a determination usually required measurements on both stressed and unstressed or bulk material.

26. What is the effect of strain-hardening on the results obtained in a nanoindentation test?

Strain-hardening does affect the shape of the load displacement curve and hence the results obtained. However, the strain-hardening exponent can be measured by using data obtained with a spherical indenter and comparing the shape of the load-displacement curve to that calculated using a theoretical model that contains the strain-hardening exponent as an input. There are other theoretical models available that allow the strain-hardening exponent to be measured from the load-displacement curve. None of these procedures have been shown to be generally applicable and they may depend upon the material being tested.

27. What is the importance of surface roughness?

Surface roughness becomes important when the mean asperity height is comparable to the depth of penetration for a particular indentation test. In most cases, a "mirror" finish is required for nanoindentation testing at penetration depths less than 100 nm. If it is not possible to polish or prepare surfaces, then, if they are bulk specimens, the indentation load can be increased (up to 500 mN or so) so that the depth of penetration is large (in the order of microns) compared to the roughness of the specimen surface. Many indentations are sometimes needed to obtain good statistics on the specimen surface. A spherical indenter is sometimes used for rough specimens at lower indentation depths, the intention being that the indenter will specimen a greater area of surface than a Berkovich or pointed indenter.

28. What is "PID feedback control"? What is it for?

PID stands for "proportional" "derivative" and "integral" methods of control. It refers to a method of feedback for controlling actuators. In some nanoindentation test instruments, PID control is used to control the application of force and also specimen positioning by the combination of sensors, actuators, and electronics circuits. For force application, some instruments operate in what is called "open loop" mode. This means that there is no force sensor in the system. Rather, the force actuator is calibrated to provide a given level of force when fed a certain current or voltage signal. Closed loop mode allows precise application and maintenance of the desired force.

In force feedback control, the signal voltage applied to the load actuator is adjusted continuously by an electronic feedback system so that the commanded load set by the user is actually applied to the indenter. This is called "closed loop" control. The inclusion of a closed loop mode of operation is very important for scratch testing, where the specimen surface is typically translated underneath the indenter. Any variations in height or any slope in the specimen surface means that while the coil current in an open loop instrument might be kept constant, the actual load applied to the indenter may vary as the mounting springs are deflected as the specimen moves. A closed loop feedback system ensures that the load applied to the indenter is kept at the commanded level and takes into account any variations in height or slope of the specimen surface during measurements.

For specimen positioning, some instruments use a stepper motor without any positioning feedback control. This means that the specimen position depends upon an accurate measure of pulses sent to the stepper motor. Errors are introduced when steps are dropped and the specimen does not move as desired. These errors are cumulative and for an array of many indentations, it will not be possible to accurately place indentations. In a closed loop system, a separate displacement sensor is used to control the motor so that the desired position is always achieved.

PID control of both force and specimen position refers to the amount of signal sent to the actuator. The "proportional" component of the actuator signal is that given by the difference between the set point and the target. For example, for a long specimen traverse, it is desirable from a point of view of efficiency, that the specimen move quickly to a point somewhere near the desired target position, and then move slowly up to the target, and if any overshoot occurs, to zero in on the target. PID position control enables this efficient mode of operation. For force application, overshooting is not desirable since then the load-displacement curve will be in error. For closed-loop force control, an over-damped system is employed with proportional control only.

29. What is the step size and range for the specimen positioning system?

It is very important for many applications of nanoindentation to be able to position the indenter and the specimen very precisely. For example, it may be necessary to make a series of indentations in a multi-phase material on inclusions that have dimensions in the order of a micron. In most nanoindentation test instruments, the specimen is moved and the indenter kept stationary. The specimen is often mounted on a platform that in turn is attached to motorized stage in X and Y directions. Non-identifying specifications for typical nanoindentation instruments are given below.

Table A6.3 Typical specifications for specimen positioning.

Instrument	#1	#2	#3	#4	#5
Range of Movement	Not available	120 × 100 mm	30 × 21 mm	Not available	Nominal 50 × 50 mm
Resolution	≈ μm	0.5 μm	0.25 μm	Not available	0.1 μm

Step size may not be the most important criteria in some applications. Although the step size might be specified at a very low level, the overall positioning accuracy depends upon the quality of the encoder used and typically, over a long traverse, the positioning accuracy is about a micron for most encoders.

30. Can nanoindentation be used to measure fracture toughness?

There are several methods of measurement for fracture toughness using an indentation technique. For brittle materials, cracks can be induced at the edges of the indentation and the crack length measured and used in an appropriate formula to give K_{IC} directly. Alternatively, for ductile materials, some workers are of the opinion that the load enclosed by the load-displacement response is a measure of fracture toughness. Such determination however ignores toughening mechanisms such as bridging and other R-curve behaviors.

31. Can nanoindentation be used for specimens with surface modified layers?

Yes, there are several methods. If the modified layer is extensive (more than a micron or two) then a sectioned specimen should be tested. If the modified layer is very thin, less than a micron, then measurements of modulus and hardness vs depth of penetration into the specimen can be obtained using two methods.

One method uses a continuous application of an oscillatory motion onto the force application during the loading sequence. The purpose of this is to measure the stiffness of the contact (dP/dh). From this information, a measure of the elastic modulus and hardness is available as a function of indentation depth. This method requires that the dynamic characteristics of the instrument be calibrated out of the measurement by oscillating the indenter in air.

Another way to obtain modulus and hardness vs depth information is called the partial unload technique. The partial unload technique is used by a number of manufacturers. In this mode, the indenter is partially unloaded after achieving each load increment and the stiffness of the contact determined in exactly the same way as for the maximum load data point.

32. For the multi-phase materials, what is the minimum distance between the indentation and the boundary of the phase to avoid influence from the neighboring phase?

The minimum specimen positioning step size is typically less than a micron with an overall accuracy of about a micron over long distances. The width of an indention for a Berkovich indenter is approximately 7 times the depth. Thus, to avoid influence from neighboring phases, the indentation should be positioned as accurately as possible and a small load used if the phases measure less than 5 μm across.

33. Can the hardness be expressed as HV?

There is a direct correspondence between the Meyer hardness H (that measured by nanoindentation test instruments) and Vickers hardness (HV). The formula is: HV = 94.495 H.

34. How can the specimen be mounted without wax? When using the wax, how can the wax be removed completely?

Specimens can be mounted with any type of glue, but some glues may be difficult to remove. Some glues will dissolve in acetone if soaked overnight. Wax bonding is a common specimen mounting method because the specimen can easily be removed from the mounting by reheating. The melting temperature of the wax is typically about 60 C. When dried, the wax can often be chipped off with a spatula and be completely removed. Some manufacturers offer a vacuum specimen mounting which uses vacuum applied through holes or slots in the mounting surface to hold flat specimens (e.g. silicon wafers) without the need for glue.

Appendix 7
Specifications for a Nanoindenter

Nanoindentation test instruments (nanoindenters) are available from a small number of specialist suppliers. Due to the low volumes of instrument sales, and the relatively high proportion of technical support required for new purchasers, the cost of this type of equipment is usually very high. Most purchases are thus the result of an open tender process. In this process, a technical specification for the supply of suitable equipment is required. To assist in this process, the following sections describe the range of issues that require addressing in such a tender process.

A7.1. Instrument Function

The system must be capable of performing experiments of the classical nano-indentation type performing instrumented loading and unloading of the indenter onto a specimen to provide indentation hardness and indentation modulus vs depth at precise positions on the specimen surface.

A7.1.1 Loading Techniques

The instrument should be software controlled for data acquisition and data analysis. The instrument should provide the following loading techniques:

1. force control
2. depth control
3. constant strain rate
4. step loading

The instrument should also offer depth profiling of material properties, i.e., a means to provide values of indentation modulus and indentation hardness as a function of depth of penetration into the specimen surface.

A7.1.2 Test Schedules and Parameters

The instrument should allow a variety of test schedules to be selected, from a simple loading and unloading, to multiple partial unloadings, to a user-specified sequence of loadings, unloadings, and hold periods. In particular, the instrument should provide the following user-selectable test parameters as a minimum requirement:

1. Initial contact load
2. Maximum load
3. Number of load increments
4. Number of unload increments
5. Time period for hold at either or both maximum load or final unload
6. Time period for delay for starting test (up to 99 hours)
7. Time period to wait after each indentation
8. Number of indentations and spacing
9. User file containing custom test schedule

The instrument should offer the capability of reloading an impression without re-zeroing the depth sensor for accumulated damage determination.

A7.1.3 Indenters

The instrument should be capable of being fitted with the following standard indenter types readily available as consumables from the instrument manufacturer.

1. Sphero-conical indenter with radius from 1 um to 200 um
2. Berkovich 3-sided pyramid indenter
3. Vickers 4-sided pyramid indenter
4. Knoop 4-sided pyramid indenter
5. Cube-corner 3-sided pyramid indenter

Each indenter should be supplied with a calibrated area function. The indenter changeover must not require a service call and be performed using ordinary tools by the operator.

A7.2 Basic Specifications and Construction

A7.2.1 Displacement Range, Resolution and Noise Floor

The instrument should offer a choice of displacement ranges. The maximum displacement range of the instrument should be 20 μm. The minimum displacement range of the instrument should be no more than 2 μm. The instrument should have a theoretical resolution of less than 0.005 nm on the lowest depth operating range. The noise floor, under good laboratory conditions on the lowest depth range should be no more than 0.05 nm.

A7.2.2 Force Range, Resolution and Noise Floor

The instrument should offer a choice of load ranges. The maximum force range of the instrument should be 500 mN for high load applications. The minimum force range offered should be no more than 50 mN for good resolution at lower loads. The instrument should have a theoretical resolution of less than 0.05 uN on the lowest load operating range. The noise floor, under good laboratory conditions on the lowest load range should be no more than 0.5 uN.

A7.2.3 Minimum Contact Load

The instrument should offer a minimum contact force of no more than 5 μN under normal laboratory conditions.

A7.2.4 Load Frame Stiffness

The load frame stiffness of the instrument should be no more than 5×10^6 Nm^{-1} to minimize the compliance error correction when testing very stiff materials.

A7.2.5 Loading Mechanism

The instrument should offer a means of working in open-loop and closed-loop force feedback servo mode. In closed-loop mode, the force achieved should be less than 1% of the requested load. In open-loop mode, the instrument should offer the facility to vary the loading rate according to the load rate, depth rate, or constant strain rate. The instrument should also offer step loading, the number and magnitude of the steps being user-selectable.

A7.2.6 Specimen Height

The measurement assembly should be adjustable so that different heights of specimen may be accommodated on the specimen stage. The maximum height of specimen that can be mounted on the motorized specimen stage should be no less than 50 mm.

A7.3 Specimen Positioning

The instrument should provide a means of aligning the indentation site with a positioning accuracy of less than ±1 μm with a step size of 0.1 μm. The positioning function should be preferably closed loop, servo controlled in operation. The field of testing for the instrument should be no less than 50 mm × 50 mm. The specimen stage should accommodate more than one specimen holder at a time. The stage should allow movement to and from the optical or AFM stations without the need for the operator to remove the specimen from the stage. The stage movement should be operable from the keyboard or mouse using computer control. The load frame should be fitted with a z-axis drive to allow for positioning of samples with different thicknesses or sample height.

A7.4 Specimen Mounting

The instrument should come equipped with a vacuum specimen mount suitable for wafers or thin specimens for certain applications. Specimen mounts must in general have a flat surface and permit specimens to be mounted using metallographic techniques. The specimen mounting must provide a rigid means of securing the specimen during testing and viewing. The specimen mounting arrangement must accommodate specimens with different heights to be placed on the specimen stage.

A7.5 Optical Microscope and Imaging

The instrument should provide a means of viewing the indentation site before and after indentation using an optical microscope with a live image that appears on the computer monitor. The image should be able to be captured and saved to computer disk in a recognized industry format. The optical microscope should provide either monochrome output or color as an option.

The optical microscope shall provide 4 turret-mounted lenses, ×5, ×10, ×20 and ×40 objectives, that can be positioned without the need for realignment of the specimen or the microscope.

A7.6 Software

The operation of the instrument shall be fully computerized with the exception of specimen loading and microscope lens selection.

The software should allow the prospective field of view to be imaged and the indentations sites selected from this image. The software should allow measurement of features (such as cracks) on the indented specimen.

The software should display in real time, the indenter XY position, current load and displacement during testing.

The software should allow more than one specimen to be tested in an automated batch sequence of tests. The different specimens should be able to be tested using a single test specification or a different test specification.

The software should allow comma-delimited export format for all data recorded.

The software should provide an automated means for correcting for initial penetration, thermal drift, instrument compliance, and indenter area function. The user should have control over the fitting parameters needs for the initial penetration correction procedure.

The software should calculate indentation modulus and hardness according to the Field and Swain, and/or Oliver and Pharr methods of analysis using a user-selectable linear, polynomial, or power law fitting procedure to the unload data. The analysis should accommodate all the types of indenters listed in A7.1.3.

The software should have the capability of reporting results according to ISO 14577 standard for instrumented indentation testing.

The software should provide a graphical output of load-penetration curves, modulus, hardness, with a user-selectable range of quantities for the X and Y axes. Multiple tests are to be able to displayed on the one graph and printed.

The software should provide a means for simulating the load-displacement curve based on user input of modulus and hardness according to a reverse application of the Field and Swain, or Oliver and Pharr methods of analysis.

The software should provide instrument diagnostics functions that permit the operator to determine if the instrument is working correctly.

The software should be able to be installed on any normal office personal computer for off-line data analysis.

The software should be supported by a web-site that offers free downloads of enhancements and fixes.

A7.7 Accessories

The instrument should offer the following options and upgrades:

1. Atomic force microscope (AFM): The AFM attachment must allow AFM imaging of the residual impression in the specimen surface without the need to reposition the specimen other than to translate it to the AFM imaging position.
2. Scratch tester: The scratch tester shall be fitted with a lateral force sensor and allow normal force with feedback. The software shall provide output in terms of friction coefficient. The scratch tester accessory shall be capable of measuring step height.
3. High temperature stage: The high temperature stage option shall provide a means of heating the specimen to greater than 300°C and measuring the load-displacement response without excessive thermal drift.

A7.8 Instrument Mounting and Isolation

The instrument shall be mounted on a sturdy frame that is isolated from mechanical vibration. The instrument itself shall be mounted on the frame using damped springs or active isolation such as an air table. The instrument shall be enclosed in an environmental chamber that offers thermal and electromagnetic shielding. The interior of the chamber shall allow active temperature control as an option.

A7.9 Delivery, Calibration and Warranty

A7.9.1 Delivery

The system should be delivered and installed so that it is in working order before acceptance by the customer.

A7.9.2 Calibration

The calibration of the instrument must be traceable to internationally recognized national standards of force and displacement. A calibration certificate shall be supplied with the instrument.

A7.9.3 Warranty

The instrument shall be covered by a warranty against faulty workmanship and materials.

A7.10 Instrument/Supplier Checklist

Price: • Does the price represent value for money? • Is the price realistic for the specifications on offer? • Are there flexible payment plans available?	
Specifications: • Do the claimed specifications stand up to close examination? • Can the supplier prove that the instrument operates at the claimed specifications consistently? • Will the supplier undertake sample tests prior to purchase? • Do the specifications meet the desired requirements? • Does the instrument require a special environment that might need to be budgeted for separately? • Can the indenter be easily changed? • Is the instrument versatile? • Does the instrument make economical use of floor space?	
Software: • Is the software easy to use? • Can it be installed off-line for data analysis? • Is it based on a well-known programming language? • Will the supplier offer free updates? • Is the software maintained by the supplier? • Will the supplier customize the software for particular applications? • Are the outputs from the software easy to export?	
Client base: • Is there a significant installed base of existing instruments? • Are instruments already installed still being used? • Are the instruments installed at recognized laboratories? • What are the opinions of existing users? • Are the results obtained from the instrument widely published?	
Supplier: • Is the supplier a reputable organization? • Can the supplier provide academic support? • Does the supplier have adequate financial resources to support the instrument? • Does the supplier have a stable group of specialists dedicated to the instrument? • Will the supplier deliver within a reasonable time period? • Does the supplier offer a substantial warranty period? • Does the supplier have a good reputation for service?	

Index

Mechanical Engineering Series *(continued from page ii)*

E.N. Kuznetsov, **Underconstrained Structural Systems**

P. Ladevèze, **Nonlinear Computational Structural Mechanics: New Approaches and Non-Incremental Methods of Calculation**

P. Ladevèze and J.-P. Pelle, **Mastering Calculations in Linear and Nonlinear Mechanics**

A. Lawrence, **Modern Inertial Technology: Navigation, Guidance, and Control, 2nd ed.**

R.A. Layton, **Principles of Analytical System Dynamics**

F.F. Ling, W.M. Lai, D.A. Lucca, **Fundamentals of Surface Mechanics: With Applications, 2nd ed.**

C.V. Madhusudana, **Thermal Contact Conductance**

D.P. Miannay, **Fracture Mechanics**

D.P. Miannay, **Time-Dependent Fracture Mechanics**

D.K. Miu, **Mechatronics: Electromechanics and Contromechanics**

D. Post, B. Han, and P. Ifju, **High Sensitivity Moiré: Experimental Analysis for Mechanics and Materials**

F.P. Rimrott, **Introductory Attitude Dynamics**

S.S. Sadhal, P.S. Ayyaswamy, and J.N. Chung, **Transport Phenomena with Drops and Bubbles**

A.A. Shabana, **Theory of Vibration: An Introduction, 2nd ed.**

A.A. Shabana, **Theory of Vibration: Discrete and Continuous Systems, 2nd ed.**